WORKSHOPS

build your skills

Easy Electronics

How to Use a Breadboard

Capturing Your Projects

Cultivating a Maker Mindset

More coming soon...!

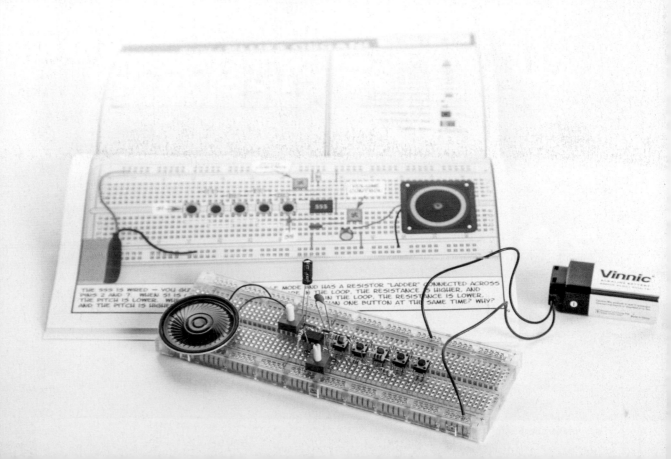

CONTENTS

30

ON THE COVER:
Naomi "SexyCyborg" Wu shows
off the sino:bit board and
its ability to display Chinese
characters.

Photo: Theodore Kaye

Courtesy of Naomi Wu, Matteo Stucchi, Adam Woodworth, Hep Svadja, Rich Nelson, Nat Heckathorn, Otto DIY

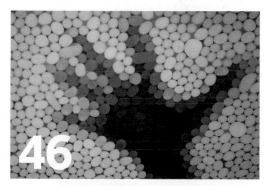

Make:

EXECUTIVE CHAIRMAN & CEO
Dale Dougherty
dale@makermedia.com

CFO & COO
Todd Sotkiewicz
todd@makermedia.com

VICE PRESIDENT
Sherry Huss
sherry@makermedia.com

EDITORIAL

EXECUTIVE EDITOR
Mike Senese
mike@makermedia.com

SENIOR EDITOR
Caleb Kraft
caleb@makermedia.com

EDITOR
Laurie Barton

MANAGING EDITOR, DIGITAL
Sophia Smith

PRODUCTION MANAGER
Craig Couden

EDITORIAL INTERN
Jordan Ramée

CONTRIBUTING EDITORS
William Gurstelle
Charles Platt
Matt Stultz

CONTRIBUTING WRITERS
Marcos Arias, Dr. Saad Biaz, Rei Betsuyaku, Nicole Catrett, Dr. Richard Chapman, Larry Cotton, DC Denison, Paloma Fautley, Shirley Feng, Brad Halsey, Harrison Fuller, Yuji Hayashi, Lin Jie aka 00, Adam Kemp, Lit Liao, Carrie Leung, Rich Nelson, Forrest M. Mims III, Andrew Morgan, Brookelynn Morris, Camilo Parra Palacio, Paul Moore, Sean Michael Ragan, Julia Skott, Violet Su, Sarah Vitak, John Wargo, Adam Woodworth, Naomi Wu, Vicky Xie, Tomofumi Yoshida

DESIGN, PHOTOGRAPHY & VIDEO

ART DIRECTOR
Juliann Brown

PHOTO EDITOR
Hep Svadja

SENIOR VIDEO PRODUCER
Tyler Winegarner

MAKEZINE.COM

ENGINEERING MANAGER
Jazmine Livingston

WEB/PRODUCT DEVELOPMENT
David Beauchamp
Bill Olson
Sarah Struck
Alicia Williams

CONTRIBUTING ARTISTS
Theodore Kaye

ONLINE CONTRIBUTORS
Akiba, Halil Aksu, Gareth Branwyn, Vivienne Byrd, Jon Christian, Chiara Cechini, Jeremy Cook, DC Denison, Stuart Deutsch, Wolfram Donat, Liam Grace-Flood, Brad Halsey, Keith Hamas, Tom Igoe, Ryan Jenkins, Zach Kaplan, Ted Kinsman, Becky LeBret, Joel Leonard, Dr. Evan Malone, Lisa Martin, Goli Mohammadi, Saba Mundlay, Jacinta Plucinski, Pete Prodoehl, Dan Schneiderman, Christine Sunu, Phillip Torrone, Ben Vagle, Shawn Van Every, Sarah Vitak, Glen Whitney, Dan Woods, Wayne Yoshida

PARTNERSHIPS & ADVERTISING
makermedia.com/contact-sales or partnerships@makezine.com

DIRECTOR OF PARTNERSHIPS & PROGRAMS
Katie D. Kunde

STRATEGIC PARTNERSHIPS
Cecily Benzon
Brigitte Mullin

DIRECTOR OF MEDIA OPERATIONS
Mara Lincoln

BOOKS

PUBLISHER
Roger Stewart

EDITOR
Patrick Di Justo

PUBLICIST
Gretchen Giles

MAKER SHARE

DIRECTOR, ONLINE OPS
Clair Whitmer

CONTENT & COMMUNITY MANAGER
Matthew A. Dalton

LEARNING EDITOR
Keith Hammond

DESIGN INTERN
Pravisti Shrestha

MAKER FAIRE

EXECUTIVE PRODUCER
Louise Glasgow

PROGRAM DIRECTOR
Sabrina Merlo

COMMERCE

PRODUCTION AND LOGISTICS MANAGER
Rob Bullington

PUBLISHED BY

MAKER MEDIA, INC.
Dale Dougherty

Copyright © 2017 Maker Media, Inc. All rights reserved. Reproduction without permission is prohibited. Printed in the USA by Schumann Printers, Inc.

Comments may be sent to:
editor@makezine.com

Visit us online:
makezine.com

Follow us:
🐦 @make @makerfaire @makershed
google.com/+make
makemagazine
makemagazine
makemagazine
twitch.tv/make
makemagazine

Manage your account online, including change of address:
makezine.com/account
866-289-8847 toll-free in U.S. and Canada
818-487-2037,
5 a.m.–5 p.m., PST
cs@readerservices.makezine.com

Issue No. 61, February/March 2018. *Make:* (ISSN 1556-2336) is published bimonthly by Maker Media, Inc. in the months of January, March, May, July, September, and November. Maker Media is located at 1700 Montgomery Street, Suite 240, San Francisco, CA 94111. SUBSCRIPTIONS: Send all subscription requests to *Make:*, P.O. Box 17046, North Hollywood, CA 91615-9588 or subscribe online at makezine.com/offer or via phone at (866) 289-8847 (U.S. and Canada); all other countries call (818) 487-2037. Subscriptions are available for $34.99 for 1 year (6 issues) in the United States; in Canada: $39.99 USD; all other countries: $50.09 USD. Periodicals Postage Paid at San Francisco, CA, and at additional mailing offices. POSTMASTER: Send address changes to *Make:*, P.O. Box 17046, North Hollywood, CA 91615-9588. Canada Post Publications Mail Agreement Number 41129568. CANADA POSTMASTER: Send address changes to: Maker Media, PO Box 456, Niagara Falls, ON L2E 6V2

CONTRIBUTORS

What do you do when you're stumped on a project?

Nicole Catrett
Rainbow Lightbox
(El Cerrito, CA)
I try to break the project down into smaller parts that I can work on one at a time, until I have it licked!

Jazmine Livingston
Dot Journaling Review (Oakland, CA)
I seek advice from folks not affiliated with it to get a new perspective. If that doesn't help, I switch gears and do something else. The answers always seem to come when you aren't seeking them!

Julia Skott
Reality Check (Sköndal, Sweden)
I will either turn to the internet, or to a friend who knows what I'm talking about and can help me talk it through. Sometimes things have to go take a time out in the Angry Cupboard, though.

Theodore Kaye
Cover photo shoot (Shenzhen, China)
Here in Shenzhen, visionary pragmatist Deng Xiaoping's maxim still guides us through problematic stumpers: "cross the river by feeling the stones."

Lit Liao
Inside Edition (Shenzhen, China)
Search for clues from the community. A diverse community is always useful to get suggestions, but one should still know there won't be perfect answer to your question most of the time.

PRINTED WITH SOY INK

Reader Builds *and* Inspired Makers

Reader **Dave Matthews** built an awesome looking "Blooming Flower Night Light" (*Make:* Vol. 59, page 62) for his friend David Holmes and family as a housewarming gift. Great job!

HURRICANE STRAIN RELIEF

I've being a subscriber for a long time, normally interested in reading "electronics-only" related articles. Hurricane Maria destroyed our island, affecting electricity, water, communications, besides others basic things, services not available for a long time. My apartment complex did not allow us to install power generators, limiting my activities to mostly reading in daylight (no cell coverage, no internet, no cable TV). During that time, I spent many hours reading previous issues of *Make:*. I discovered I was missing "half the fun" reading back to those non-electronics related articles. I found that everything correlates to the real hearth of true makers.

Keep doing such an excellent magazine. I won't miss any part of it.
–Edgar Polanco, San Juan, Puerto Rico

Make: **Executive Editor Mike Senese responds:**

Edgar, first, let us express our sincere concerns and hopes for Puerto Rico to recover as quickly and thoroughly as possible. With that, thank you for sharing your rediscovery of the inspiration that can be found away from the distractions of our electronic devices. We're happy to be a part of that for you. Please let us know what you end up putting together!

MEETING YOUR HEROES AT MAKER FAIRE

My name is Diego Chavez, I am a 16-year-old maker from Guatemala, and I write to you to thank you. Four years ago I started having an interest in making, and it didn't take me long to get into YouTube and find Kipkay's weekend projects and many other *Make:* videos that helped me learn and start making my own projects. When I was 13, for my birthday I asked to go to Maker Faire Bay Area, and I can only describe it as one of the best experiences of my life. I was able to see some crazy things that I never would have imagined could be made. I was able to meet my childhood hero Adam Savage and take a picture with him right after I listened to an amazing speech he gave. It is an understatement to say that day shaped me; it gave me the motivation and the drive to keep learning. *–Diego Chavez, via email* ●

Prison Ban of the Month

» **LOCATION:** Pennsylvania Department of Corrections

» **TITLE:** *Make:* Vol. 59

» **REASON:** "Information contained on pages 46–48; 58–61 provide information regarding secret cabinet locks and DIY thermal imager kits."

pennsylvania

To: Maker Media, Inc.
1700 Montgomery St., Suite 240
San Francisco, CA 94111

From: Office of Policy, Grants & Legislative Affairs
Pennsylvania Department of Corrections
1920 Technology Parkway
Mechanicsburg, PA 17050

Date: September 26, 2017

RE: Publication Denial

Please be advised that the following publication has been denied to all inmates housed in the Pennsylvania Department of Corrections:

Make, v. 59, Oct/Nov 2017 – Home Hacks

The publication was denied for the following reasons:

Information contained on pages 46–48; 58–61 provide information regarding secret cabinet locks and DIY thermal imager kits.

If you would like to appeal this decision, you have 15 working days to set forth the reasons why the decision to deny the publication was erroneous. Responses must be sent to:

Office of Policy, Grants and Legislative Affairs
PA Department of Corrections
1920 Technology Parkway
Mechanicsburg, PA 17050
ra-crdocpolicy@pa.gov.

To review PADOC publication policies, please see www.cor.pa.gov.

Policy Office | 1920 Technology Parkway | Mechanicsburg, PA 17055 | 717-728-2573 | www.pa.gov

David Holmes

Make: Amends

In *Make:* **Vol. 60 we forgot to credit Rob Nance's excellent exploded diagram for "iPad Teleprompter" on page 84. Sorry Rob!**

Culture and Creativity

BY MIKE SENESE, executive editor of *Make:* magazine

This rotating cardboard shawarma automata is a student project from the first week of the Sabic camp.

Sarah Dosary shows off TekSpacy's collection of *Make:* magazines.

Part of my job is traveling to Maker Faires and other *Make:* events, sometimes in far-off places, where I get to meet the diverse and enthusiastic members of our community, see the projects they're making, and help share their stories.

Last year I had two big trips: the first to Maker Faire Kuwait, which proudly hosted a large number of women engineers as exhibitors, something I documented in *Make:* Vol. 58 ("Standing Out," page 18). The second was a 12-day trip to Riyadh, Saudi Arabia for the first segment of a 3-week maker-education summer camp at industrial conglomerate Sabic, for which we designed and ran the curriculum.

I embarked on that trip with a lot of curiosity and a little trepidation. What I found there was, in many ways, not unlike most other places I've been — the hosts were great, the locals were welcoming, and the facilitators we worked with, recent grads from Saudi Arabia and the surrounding countries, were a warm and open group, instantly bonding and quickly grasping the content.

As is customary in Saudi Arabia, the program had the boys' and girls' groups divided and placed into two different parts of the facility, with facilitators of the same gender, and a strict rule that the men couldn't enter the women's side whatsoever while the students were present. However, after the students left, the facilitators would convene in the girls' classroom to discuss each day's progress. In those meetings we all quickly noted that, day after day, the girls' projects exhibited creativity and quality that exceeded that of the boys' — in spite of (or perhaps due to) the limitations placed on them.

The highlight of the trip was Sarah Dosary, one of the facilitators, inviting us to her women-only makerspace, TekSpacy. We drove across Riyadh in the heat of the afternoon sun to get to her pristine, cozy shop, fully equipped with 3D printers, laser and CNC cutters, design tools, and more. She explained that women could work there without their niqabs and abayas, focusing on making projects and starting businesses. It stood in stark contrast to what we hear about how women in Saudi Arabia live. While unable to drive cars and requiring a male escort for many day-to-day activities, here women could work with business incubators to start companies and help other women do the same. Within the confines of the Saudi society, Sarah had found a way to express herself, not unlike the girls in the class she was helping facilitate.

In this issue, I'm excited for our spotlight on the women shaping Shenzhen. There are makers like them and Sarah all over — we would love to learn about them all and share their stories, as we aim to better promote the great diversity of our community. Please send any suggestions my way: mike@makermedia.com. ◗

Mike Senese

MADE ON EARTH

Backyard builds from around the globe

Know a project that would be perfect for Made on Earth?
Let us know: *makezine.com/contribute*

SWEET SCENES

INSTAGRAM.COM/IDOLCIDIGULLIVER

When your mom told you not to play with your food, she'd clearly never heard of *I Dolci di Gulliver*. I Dolci di Gulliver (Gulliver's Sweets) is the magical project of Italian pastry chef **Matteo Stucchi**.

Stucchi's special ingredient is a heaping spoonful of whimsy and imagination. Each dessert is part of a miniature scene. The people in his tiny imaginary land of Lilliput farm in croissants, ride in cotton candy hot air balloons, and paint layers of fruit onto a popsicle. His first shot, taken in July 2016, features three miniature whitewater rafters riding a chocolatey wave out of a molten lava cake. "Who has never dreamed of being able to raft in a chocolate river? Sometimes dreams can become reality," reads the caption. Most of the miniatures are purchased at a special store, but many of the larger objects Stucchi makes himself.

It usually takes Stucchi two or three hours to make the dessert, set the scene, and take the shot. Afterward, he gets to enjoy eating the sweets with his family. He finds inspiration in the dessert itself. "It suggests the idea because of its shape or preparation," he explains.

One of the most beautiful aspects of his work is the way Stucchi creates a sense of motion and activity in all of his scenes. Some capture suspended fruit flying through the air, or depict the middle of an explosive rocket launch. Others catch milk being spilled or powdered sugar being dumped. He tries to add surprising elements to each of his creations. Many of his pieces tell a comical story about how the people in Lilliput make these giant confections. "If you have fun creating each photo then you will also be able to excite. It fills me with joy to know that my creations excite those who watch them, it means that I have reached the goal." —*Sarah Vitak*

Matteo Stucchi

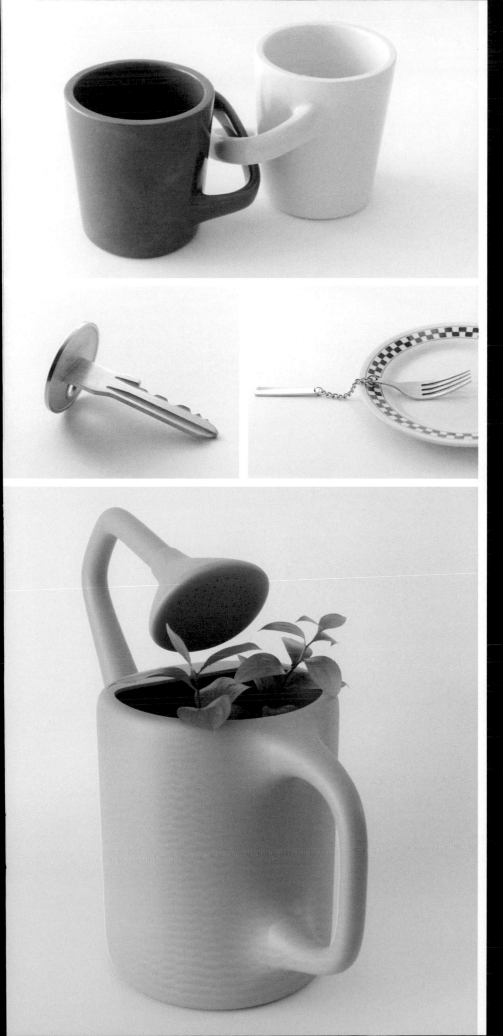

Katerina Kamprani

DYSFUNCTIONAL DESIGNS

THEUNCOMFORTABLE.COM
Good design is often taken for granted — it's the terribly designed products that call attention to themselves. Athens-based architect and artist **Katerina Kamprani** takes this a step further. Since 2011, she's been designing household objects that are ingeniously awful.

For each piece, Kamprani chooses a well-known, simple object, and analyzes it, searching for ways to "sabotage simple steps in the interaction with it." When she finds an idea that makes her laugh, she models it in a 3D program. It wasn't until much later, when she was invited to do a solo exhibition at the 2017 Zagreb Design Week, that she figured she ought to build physical prototypes. She 3D printed resin parts, carved wax for lost wax casting, sculpted clay, and collaborated with other craftspeople in order to bring her different prototypes to life.

After 10 years of working as an architect, Kamprani wanted to pivot to a more "playful" career, so she embarked on a graduate program in industrial design. Kamprani never finished the program. "After that," she says, "I had several other failures in my effort to change my career and quit architecture. It was only when I stopped trying and relaxed into the idea that my job is not my artistic playground that I started my own personal project in a very organic way. It does hurt to fail, but it also brings you closer to what suits your needs better — I needed to design for fun!"

Kamprani recently made an uncomfortable table for a group exhibition in Athens. The project happened organically, so she's not sure where it will lead her next, but she's thinking about producing objects for sale, putting them in a book, or doing creative activities with kids. —*Sophia Smith*

Peter Coffin

MADE ON EARTH

TREMENDOUS TROUSERS

PETERCOFFINSTUDIO.COM

The 450-year-old Wånas Castle, nestled in southern Sweden, is now home to an international contemporary sculpture park that has showcased artists such as Yoko Ono, Jenny Holzer, and Marina Abramovic. Brooklyn-based artist **Peter Coffin** exhibited an art installation there in 2007 that indirectly led him to create his *Untitled (Tree Pants)* series.

Coffin says he doesn't take his art too seriously, and that the project actually started as a joke while he was preparing to install some large-scale artworks at Wånas. "I found a giant pair of pants," he explains matter-of-factly, "and I decided to put them on a tree." Coffin let people discover it, and the denim-clad tree had such a positive reception that the Wånas sculpture park asked him to install more in the woods surrounding the castle. He dressed a dozen trees, tucked away for people to seek out. "For some people it became a fun experience to find them all," he says.

To place them, Coffin cut the seams and sewed the pants around the trees directly. He used a boom lift to get as high as 70 feet off the ground in order to do the installations. Levi Strauss sent him additional denim for the project, and Coffin worked with professional seamsters to sew the pants on.

Since the original series was completed, some of the pants have been reinstalled by collectors all over the world, in California, Brazil, France, Germany, and elsewhere.

So why pants? Coffin says, "It sheds light on the funny habit we have of personifying inanimate things — it's sort of our way of relating to things ... and that tells us a little bit about ourselves."

—*Sophia Smith*

Matt Holmes

A MARVELOUS MIX
HOLMESYLOGIC.CO.UK
MAKERSHARE.COM/PROJECTS/WALKMAN-COFFEE-TABLE

Matt Holmes fell in love with the design for the Walkman after watching Star-Lord dance with the portable cassette player in the first *Guardians of the Galaxy*. Before he even finished watching the film, Holmes knew that he needed to have one. "After some fruitless searches on eBay and the like, I discovered how much they actually cost," Holmes said. "Once I'd recovered consciousness, I resolved to build my own."

He made a couple of models, but still didn't feel like he was truly finished. He knew the concept of the Walkman could be expanded upon, but how? As fate would have it, Holmes' new apartment was in dire need of a coffee table, so he decided to

supersize the Walkman.

"I found some very good reference pictures online, which together with a few assumptions allowed me to draw up some simple plans," Holmes said. "Personally, I tend not to think the whole build through before I begin. I like to have a rough idea of my general plan of attack and then figure it out as I go."

The main body of the table is made from medium-density fiberboard (MDF). Holmes cut the Walkman's door with a thin jigsaw blade and made the giant cassette tape with a combination of laser-cut MDF and clear acrylic. The volume sliders, mic grille, and other small parts are 3D printed, which was

ultimately faster and higher quality than making them by hand.

Holmes admits that the hardest part was painting the table to look like a Walkman. "The biggest issue I had was getting an acceptable finish on the paintwork. Spraying such a large item with spray cans meant it was hard to not have streaks in the final finish," he says.

Although he's currently satisfied with it, Holmes says he's not completely done with the table. He'd like to give it the ability to play music, either via Bluetooth or by installing a vinyl record player inside.

—*Jordan Ramée*

FIGHTING DISASTERS

Written by Brad Halsey

FEMA/K.C. Wilsey

The extreme circumstances surrounding humanitarian relief require **extreme making**

BRAD HALSEY is the CEO of Building Momentum. He loves problem solving in disaster/conflict zones, thrives on teaching people how to make, is a musician turned chemist turned maker, and is constantly in awe of his wife and three kids.

THE first thing the Field Ready team, a nonprofit humanitarian organization, saw when we got off the plane in St. Thomas this past October was the utter devastation left behind by two Category 5 hurricanes. The problem was almost too big to comprehend. How does the NGO I volunteer with even begin to dig in and make a difference? Well, sort of like peeling an orange, you just jam your thumb in and get after it.

I learned to do this in an extreme place. Corkscrewing in a dilapidated ex-Soviet prop plane from 20,000 feet straight down into Baghdad was my first, albeit terrifying, glimpse of how problem solving is almost entirely about understanding the situation. The reason we dropped out of the sky right over the airport was that the insurgency had been shooting down aircraft. The sketchy plane I was crammed into belonged to a small commercial outfit that found a niche market getting non-military personnel into Iraq. Because they didn't have access to all of the fancy missile countermeasures they would just spiral straight down to the airfield from a height known to be out of the range of insurgent missiles. Fastest (and most gut-wrenching) airport approach ever — and a great example of problem solving.

From that moment in Iraq to my recent disaster relief endeavors in the Caribbean, I began formulating and refining the process for humanitarian problem solving and making. It is an approach I am comfortable with, and one that is being pioneered by Field Ready worldwide, allowing me to develop a rough and totally non-comprehensive loose set of guidelines that might help others peel the orange.

These guidelines were initially born from my time creating solutions to problems in a conflict zone. The similarities between man-made and so-called natural disasters are striking. Turns out, there are really only three things you need to do to successfully effect change in these areas:

1. Roll in it.
2. Make it/break it.
3. Deliver. Fast.

Although I listed these things in order, reality dictates they happen somewhat concurrently. At a minimum this process is iterative, moves fast, and gets intertwined quickly.

ROLL IN IT

As I ran around Iraq trying to help as the Army's "embedded geek" it became very apparent to me that in order to successfully problem solve with making on-the-ground technology solutions I needed to immerse myself in the world of the soldier. I went on missions to witness the issues firsthand. I saw the existing tech and infrastructure they had to work with and used that as a starting point. I felt the odd juxtaposition of terror and monotony as they did and rolled in all of it — the sweat, dirt, fear, unknown, sweltering port-a-potties, and existing equipment — to help frame the problem's faced.

Disaster response is no different in approach. The enemy is less resolved but potentially no less dangerous. The problems can be blurrier. The immense destruction is overwhelming and finding a starting point is seemingly impossible. But, again, at some point you just jam your thumb into it. You get off the airplane and find a place/group/organization that is doing something related to building or making solutions and jump in.

Recently, in the Caribbean, the Field Ready team found a local workshop (My Brother's Workshop in St. Thomas) that was trying to repair roofs on a few houses among other things. We jumped in to assist and interact with the community and their problems. We rolled in it. And that led to identification of other problems — and then projects — that we could make

Aerial views of hurricane damage in St. Thomas

Workers install a temporary roof on a hurricane damaged home in St. Thomas

An American flag flies high over burned homes in the Waldo Canyon Fire

St. Thomas residents clearing drains to stop flooding

Tsunami damage in American Samoa

and deliver. This all occurred in the first 24 hours on the island and generated the momentum to create several tech solutions during my week there.

Immersion must be with eyes and ears open and brain firing, but also mouth moving. Communicating ideas and the understanding of the problem to others — whether they are soldiers, disaster victims, or colleagues — concisely and frequently helps in the rapid prototyping process, even if the prototype is designed, developed, tested, and subsequently killed all in your head.

Not everyone can go to conflict/disaster zones, nor does everyone need to. Problems and solutions can be worked at home as long as the environment can be simulated and understood as much as logistically and feasibly possible. Losing sight (i.e. immersive understanding) of what is helpful to the person operating the technology results in cool gadgets that won't get used (e.g. Google Glass).

An anecdote of this exact myopia involved an Army project that was initially designed to help the overtasked electrical system of the brand-new armored vehicles that were being shipped to Iraq and Afghanistan. As soldiers needed more tech on missions the demand for vehicle power skyrocketed. Brownouts and even blackouts in vehicles were becoming common. The Army decided to rapidly prototype an improved power distribution system that could augment the strained infrastructure of each vehicle. But as development time wore on people added more capabilities, more features, more improvements, and more scope-creep. Fast forward to the end, to the dismay of the solider, the systems were never fielded as the Army optimized them to death. Remember, perfect is the enemy of good, especially in conflict and disaster zones.

MAKE IT/BREAK IT

Don't roll in it for too long! The environment is always changing. The enemy is adapting. The food is dwindling. Get started making! For many makers there is a tendency to learn as much as possible about the problem before getting to work. More information reduces build stress and makes the development process more comfortable, right? Well, prepare to be uncomfortable.

I never had the entire picture or all the parameters when I was in the thick of it. At some point I just made something. It likely sucked. I threw it out and tried again. I would break all sorts of tools and tech in the process. I have burned up whole laboratories, power supplies, furniture, computers, expensive motors, and most power tools. I have destroyed nearly every piece of tech I have used at least once. But that's how I do it, and unapologetically.

But really it's about fear. Not necessarily the fear of breaking something or screwing it up, but the fear of looking like a failure to people who need me to solve problems. The fear of feeling like I just can't learn the tech. I am supposed to be the professional problem solver, right? And yet sometimes I squeeze out failures (what I call "tech-turds"). This can't be how extreme making is done, right?

Unfortunately — yes. That's how you quickly get to the solution. This is probably not a great process for learning to land a plane; with that, iteration past the first time is a waste (or far worse). But when you need

> **Problems in disaster areas:**
> power (for lights, cooking, refrigeration, water pumps, AC); clean water/fresh water; instant traffic control (ad hoc traffic lights); communications (when cell towers are knocked out); fixing existing equipment — making replacement parts; extraction of victims (for conflict and disaster zones — making litters, medical equipment infrastructure, ad hoc fixes for emerging problems)

to make tech solutions iteration may be your best friend. Also, the people you are trying to help have a high risk-tolerance. They are desperate for help. You really don't have the time to be afraid of digging in or learning a new tool, just do it. It's actually kind of a luxury to not have time to think whether you should or shouldn't try something ... perhaps a mindset we all should adopt more often (but that is likely the topic of another article).

DELIVER. FAST.

I have had the privilege of training nearly 200 Ph.D. scientists, engineers, teachers, soldiers, Marines, and professionals in extreme making and I tell them that the single most important thing you can do when making a solution for someone is to deliver it. Show up and deliver — it's the single most important thing you can do. I even said it twice for effect. It sounds obvious but doing it is hard. And sometimes damn hard.

One evening in Baghdad I received an email from a unit that was constantly taking fire and getting blown up by IEDs. The enemy would hide in alleyways and fire RPGs at

vehicles as they passed. Our unit needed a camera solution that could see way in front of their vehicle. A camera in front could help identify these pre-positioned insurgents and even possibly spot IEDs hidden around objects on the road. That's all the info I had to work with but it was enough to get started.

To help solve this problem, I worked all through the night on a pan/tilt and zoom camera system that could be mounted on an already existing pole. I scheduled a helicopter the next afternoon to deliver it. I showed up that evening with a crappy but workable prototype. They were astounded that I had received the email just 24 hours ago and was at their doorstep, gunfire and rockets aside, with ... well ... something passable as a tech solution. I installed the system at first light and went out on a mission with them that day to ensure it worked. And although we didn't catch any insurgents or IEDs (despite plenty of small arms fire), I proved to them I could be counted on to deliver on short notice. That trust sprouted several iterations of the camera system as well as other projects that helped a unit

in a dire situation.

Delivering quickly might prove to be even more important in disaster recovery scenarios where the population is already suffering, exhausted, and feeling abandoned. Talking and working with survivors only becomes fruitful when you return the next day with a solution, however jury-rigged. Doing so makes the community feel like they are a part of the rebuilding process even if it's through the work of someone else.

This is exactly what Field Ready did in St. Thomas and in other places ranging from Nepal to Syria. We were able to roll in and work with a great partner. Quickly finding workable problems to solve, we made useful things and delivered them, fast.

The important lesson is this: all of that rolling in it and iterative making is for naught if you don't show up quickly with something in hand. Not in a month — tomorrow. Extreme making demands this kind of turnaround. Because if you can't deliver quickly, the people in need will lose faith in your ability to help them. You'll quickly become a marginalized battlefield or disaster zone tinkerer. And that's not cool.

Not all solutions can be made this way, or this quickly. But I think elements of this rapid problem solving process should be interwoven into all making and creating. The world needs to see and experience your ideas. Life is short. Make something important. ◑

HoloKit

REALITY CHECK⊙

Take VR into your own hands with these DIY builds

Written by Julia Skott

JULIA SKOTT
is a journalist, writer, podcaster, and potter who also knits and tries to keep plants alive. She tweets in Swedish, English, and bad puns as @juliaskott.

VIRTUAL AND AUGMENTED REALITY MAY FEEL LIKE THE DOMAIN OF BIG COMPANIES AND GAMING GIANTS, but it should come as no surprise that they are also playgrounds for all sorts of DIY endeavors. Whatever kind of reality you like, here are some interesting projects to play around with.

GETTING NOWHERE FAST

Remember when virtual reality involved some sort of weird structure that confined a player inside a little frame on a moving platform? Well, that's still a thing — and you can build that omnidirectional "**treadmill**" yourself, to make gaming super-immersive. It's definitely a big build, but easier than a lot of other VR DIY. Basically, it's a concave octagon clad with carpet, a support frame that can include bungee cords crisscrossed to keep you centered, and shoes with furniture carpet sliders to let you glide-run in place in your little bowl while you chase bad guys or gold coins or whatever

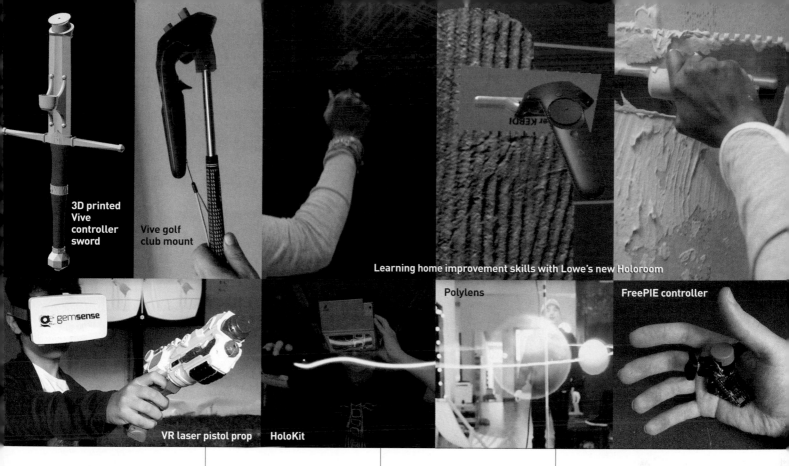

3D printed Vive controller sword

Vive golf club mount

Learning home improvement skills with Lowe's new Holoroom

gemsense

Polylens

FreePIE controller

VR laser pistol prop **HoloKit**

HoloKit - holokit.io, Lowe's, Sabba Keynejad, Florian Maurer

it is you're after. Plus some kind of sensor to keep track of where you're going. Just imagine the feeling of running around in your own living room! Details: youtube.com/watch?v=oi5DU2JfRhU.

But what about other immersive experiences? Sword fighting in VR may look real, but unless you take things into your own hands, so to speak, your Vive controller will break you out of even the most realistic epic battle. For that, FredMF designed an ingenious 3D printed sword prop controller (thingiverse.com/thing:1802871) — it even has threaded rod inside to give the right heft.

One of the harder parts of VR marksmanship games is keeping your hands steady while aiming. This adorable build from Gemsense combines an off-the-shelf laser pistol with a Blue Amber Gem Board for the space pistol of your dreams on Google Cardboard: gemsense.cool/make-your-own-vr-gun.

Similarly, the right prop can really make a game of VR golf. Dwooder's Vive golf club mount build (thingiverse.com/thing:2071035) makes clever use of an old club handle, so you can keep your swing in line while you work the back nine.

VIRTUAL TRAINING

Learning by doing is usually a good thing, but sometimes you don't want to learn by messing up your new shower or drilling into important pipes and cords. An obvious solution is to turn to a virtual room where you can practice without real consequences, which is what Lowe's new Holoroom How To offers. In-store boxes let you strap on a headset and grab a controller with some smart haptic feedback that will almost have you believing that virtual paint brush is doing actual damage — that is to say, painting. Or drilling, or performing the perfect tile job. Most people will learn better by getting to try something out than just reading instructions.

It may not be as cool as Matrix-style martial arts training, but it will likely be a lot more useful in the long run. Depending on your lifestyle.

HOLO, IT'S ME

Microsoft's HoloLens is basically augmented reality but with fancy mirrors to make the images float in front of your eyes rather than on a screen. And those fancy mirrors basically scream a challenge to make your own, right? You can find different setups online to turn your phone into a personal hologram projector, "help me Obi Wan Kenobi"-style, without the multi-thousand-dollar price tag of the HoloLens. Take some inspiration from Google's Cardboard VR — make your own from scratch using craft knives and electrical tape (instructables.com/id/Cardboard-Hololens) or buy a simple HoloKit for $35 — and some, like the Polylens, are even sleeker and more sci-fi. It does take a bit of experimentation and finagling, and most likely some programming, either way, but

it's exponentially cheaper and probably a lot more fun.

VIVE LA MANO!

HTC Vive controllers will set you back about a hundred dollars a pop, and are still just basically a stick you hold in your hand. (A cool stick, and with a doodad on the end, but a stick.) Why not take it up a notch and build a small controller that actually straps on to your hand and does all the same stuff without making you hold it?

The build, detailed at imgur.com/a/obgP6, uses FreePIE software to merge the position and rotation sensor data, simulating an Arduino joystick in place of the Vive controller. A tiny control stick sits next to the thumb, and buttons and triggers can be placed for access by relevant fingers in a similar fashion — or however you prefer and have the patience to program. And you can still type, check your Instagram, or do whatever else you want to do, without putting down your controller. ◑

Tipping the Scale

Scott N. Miller talks prototype to production, manufacturing insights, and Dragon Innovation's next moves

Scott N. Miller

Some of the products that Dragon Innovation has supported: the Jibo social robot (above), the Pebble smartwatch (top right), and the Petnet SmartFeeder (middle and bottom right).

Jibo

Long before Make:, or Kickstarter, or Arduino, mechanical engineer **Scott N. Miller** was working on the hardware frontier. In the late '90s and early '00s, he led iRobot's Roomba technical team. He lived in China for four years and scaled the robotic floor cleaner from functional prototype to a three million unit production run. Miller co-founded Dragon Innovation in 2009, which has guided over 300 companies — including MakerBot, Pebble, Ring, and Formlabs — through the manufacturing process. In August 2017, Dragon Innovation was purchased by the technology distributor Avnet. More recently, Kickstarter, Avnet, and Dragon launched Hardware Studio, which offers a free toolkit as well as a personalized support program for selected groups of Kickstarter creators.

What manufacturing terms should Maker Pros know?
Cost, schedule, and quality.

» Cost has two components: the cost of goods sold and the fixed cost, which most people forget. First-time hardware entrepreneurs will often take the cost of their materials, add it up, and say, "Okay, that's how much money I need." But that leaves out all the fixed costs: things like tools, injection molds, fixtures, stencils, FCC compliance, labor, scrap, overhead, profit for the factory ... there's a long list of unknown unknowns.

DC DENISON is the co-editor of the *Maker Pro Newsletter*, which covers the intersection of makers and business, and is the senior editor, technology at Acquia.

» For the **schedule**, you must know the key milestones: factory selection, handover, engineering pilots, and so on. Then, having a rough idea of what the duration is between each step. Make sure you don't miss the Christmas season, where so much consumer activity takes place. Don't forget the Chinese New Year, when everything comes to a stop. If you end up on the wrong side of it, you're in trouble.

» **Quality** is always neglected, but ultimately it can have the biggest impact. Everybody thinks about cost first, to figure out what their margins are. And then schedule. Quality is fitness for use. Really understand how the customer is going to use your product and make sure that it's good for that use. But there are less obvious things, like transportation testing. Is it going to survive a long trip on a bumpy Chinese road, and then get loaded onto a container ship, sit in 100°F heat for four weeks, and then a flight to Miami? Then there's all sorts of abuse testings, like drop testing.

So that's the reason behind the new collaboration with Avnet, Dragon, and Kickstarter.
Yes. The challenge of manufacturing is that the early decisions are so critical, but it's difficult to get visibility into what you need to decide when.

What developments are changing the equation for hardware entrepreneurs?
One encouraging trend is a new stack of software tools for companies building hardware. Pretty much everybody starts by using Microsoft Excel for their Bill of Materials (BOM). As companies get into higher volume, they look for things that are more capable, but there's not a lot of offerings yet. They're either forced to stick with Excel or go with enterprise-level software, which is very sophisticated, and expensive, and difficult to learn.

When we ended up putting an enterprise-level PLM and ERP system in at iRobot, we had to hire people to actually manage it, we had to pay license fees, and maintenance. That's worth it if you're manufacturing hundreds of thousands of units, but if you're growing, that's the last thing you want.

Dragon is also working on this kind of software now, right?
Yes, Product Planner. A few years ago, we introduced the Dragon Standard BOM to give the hardware community a professional grade BOM template. Product Planner also uses a BOM-centric approach, but it's much more than just a spreadsheet. We've always said that we want to give our clients the knowledge of how to "go fishing," and Product Planner is the fishing pole. You can have tools with innate knowledge in them that will guide you from here to there, but you don't need

to pay a fortune, sit in week-long training courses, or hire people to run it.

Has anything changed between manufacturing in China versus in the U.S.?
What we've seen is that there's more opportunity in the U.S. to build lower volumes. It's relatively easy to build, and raise money, if you're building 100 units; and it's easy to build and raise money if you're building 100,000 units. The hard part is building from 1,000 to 5,000 — it's the hardware "Valley of Death." I think the U.S. is situated to help address that. Nobody in China is going to be interested in building 5,000 units, and it's so much work to find a factory and deal with the language and time barriers.

Whereas if you're in Boston and you drive to a small factory in Worcester, you can sit down with somebody who speaks the same language and say, "This is what I want to build." That also gets rid of the 5 weeks it takes to ship a product across the ocean.

Are there sectors that you think are overdue for Kickstarter projects?
People have to eat, so there's a constant stream of revenue in food. I love Blue Apron, but it's a lot of chopping. So maybe there are ways to combine robotics and food to create a healthy product. There are a bunch of venture-based companies working on that, but I haven't seen any Kickstarters. ◉

Pebble, Petnet

3D Printed
Micro FPV
Quadcopter

This flyweight racer is fast, first-person, and fearless fun to fly

Written by Adam Kemp

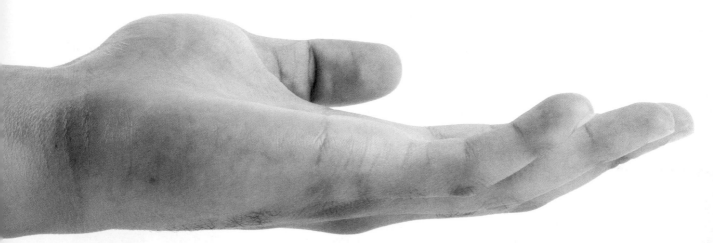

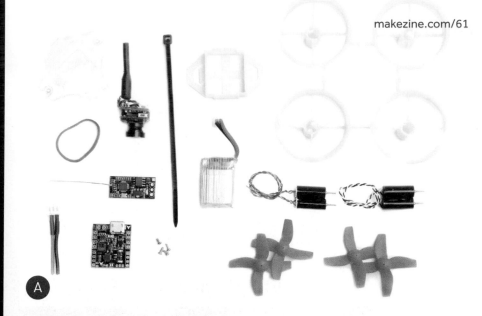

TIME REQUIRED:
30–60 Minutes
COST:
$60–$80

MATERIALS
FOR THE MICRO QUADCOPTER:
» **3D printed parts: frame (1) and control box (1)** Download the 3D models free from thingiverse.com/thing:2037157.
» **Motors, 19000KV, 6mm×15mm: clockwise (2) and counterclockwise (2)** Get the 4-pack from Crazepony, Amazon #B01N9ETA9Z amazon.com.
» **Propellers, 31mm diameter, 0.8mm mount: CW (2) and CCW (2)** Crazepony 8-pack, Amazon #B01MZ3UZGQ
» **Camera, AIO type with canopy, 5.8GHz transmitter** Amazon #B01LYAW6S6
» **Battery, LiPo, 3.7V 150mAh** for E010 or H36 copter, such as Banggood #1075947, banggood.com
» **Screws, M1.2×3mm self tapping (4)** Amazon #B01M1CI4PA
» **Rubber bands, ~1cm diameter (2)** or hair elastics
» **Receiver, 6-channel PPM, FlySky FSI6-RX compatible** I used the Usmile Tiny 6CH, Amazon #B01MXH8XQ1.
» **Flight controller, Eachine Tiny F3 Brushed** Banggood #1087648
» **Cable tie, ⅛", 18lb strength** aka zip tie
» **Power connector, male JST-PH jack** Banggood #1147298
» **Wire wrapping wire, 30 gauge, insulated**
ALSO NEEDED:
» **FPV goggles, 5.8GHz**
» **R/C transmitter, FlySky-compatible 6+ channel**

TOOLS
» **3D printer, fused filament type (optional)** You can print the parts yourself, or send the 3D files to a service for printing. See makezine.com/where-to-get-digital-fabrication-tool-access to find a printer or service you can use.
» **Soldering iron** adjustable temperature recommended
» **Solder, rosin core**
» **Snips or flush cutters, small**
» **Wire strippers, 30 gauge ideal**
» **Screwdriver, #1 Phillips**
» **Hobby knife**
» **Tape, double-sided foam squares**

ADAM KEMP teaches high school engineering and co-chairs the Science, Technology, Engineering, Art, and Mathematics department at the Princeton International School of Mathematics and Science. His book *The Makerspace Workbench* has served as a resource for thousands. Get it at makershed.com.

A

B

C

A lot has changed since I began building radio-controlled models as a kid. I started with a small wooden boat kit propelled by a Cox 0.49 motor, steered with a single servo, and controlled with a terrible AM radio. The range was very limited and the boat was slow, but man, I was hooked.

Fast-forward to today and the changes are staggering. The advent of MEMS sensors and autopilots for inertial sensing and stability control, high-efficiency motors, and multi-channel digital radios with receivers no larger than a penny have enabled us to construct R/C vehicles that once we could only imagine.

While the tiny quadcopter I'll show you how to build is not the most expensive, or complicated, and some may even scoff at its lack of brushless motors, it is by far the most fun flying experience I have ever had. The moment you power up this quad and don your goggles, you step into the cockpit of an acrobatic flying machine. The experience is straight out of science fiction yet the world you are flying through is completely real and feels 10x larger than life. So, gather all the parts you need (Figure Ⓐ), a FlySky-compatible 6+ channel transmitter, and a pair of 5.8GHz FPV goggles, and let's get started!

1. BUILD THE FRAME
The advantage of this project over commercial micro quads is its low weight, <25g, and its ability to be repaired at home for very little cost. It takes only 45 minutes for my printer to make new parts (Figure Ⓑ), so I always have a few extra on hand for when a stunt goes bad or a tree gets in the way. Start this project by visiting my Thingiverse page and downloading the latest files for the frame and control box. Print using the following settings:

Plastic Type	PLA or ABS, other plastics are untested
Layer Height	0.1mm
Infill	No less than 50%
Support	Touching build plate
Brim	No, unless you have adhesion issues

I've had a lot of success printing frames out of PLA. It is slightly more rigid than ABS, which is good for stability, but unfortunately it is more brittle. I have also attempted to use Tough Resin in an SLA-type printer, but achieved poor results as the wall thicknesses of the frame are very thin and tend to warp while printing.

Once your print is complete, remove the parts from the build plate and use your snips and hobby knife to clean up any excess plastic and supports.

2. ASSEMBLE THE ELECTRONICS
Cut, strip, and solder three 2cm lengths of wire-wrap wire to the receiver's +5V power, GND, and PPM pads (Figure Ⓒ). Depending

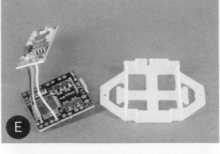

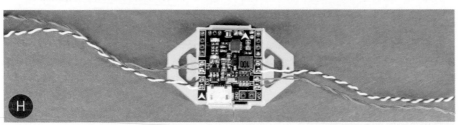

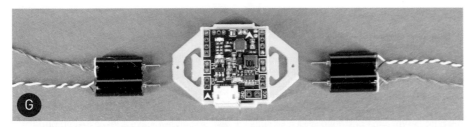

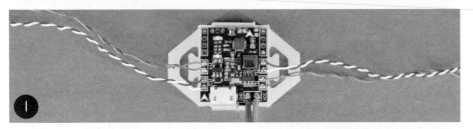

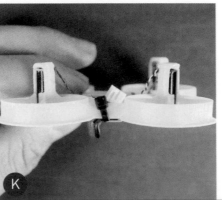

on the tiny FlySky receiver you purchase, it may or may not have a Bind button. If it doesn't, you can bind to the receiver by temporarily connecting a forth wire between the Bind and GND pads during power up. I successfully used a breadboard jumper wire to bind mine without soldering.

Strip 1mm from the three free ends of the wires on your receiver and solder them to the corresponding pads on the bottom of the flight controller (Figure **D**). Follow the drawing found on the webpage for Banggood part #1087648 to make sure you solder to the correct pads (PPM to RX2, +5V to +5V, GND to GND).

Install the receiver and flight controller (Figure **E**) into the control box by first gently pressing the receiver in place so that the surface mount electronics face the bottom of the box and the antenna slides into the notch. Fold the wires neatly between the two devices and gently press in the flight controller (Figure **F**). There should now be enough room on the bottom of the control box to access the Bind button/pad.

Gather your motors and carefully twist the motor wires together (Figure **G**). This helps reduce EMF interference and prevents the wires from getting caught during flight. If your motors came equipped with micro JST connectors, cut them off close to the connector and strip about 1mm of insulation from the end before soldering.

Tin the ends of the motor wires and solder them to the flight controller exactly as shown in Figure **H**. It's important

Adam Kemp

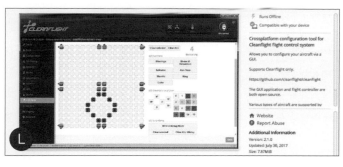

that you follow the color order as this dictates the direction the motor will spin. Two of the motors are designed to rotate clockwise (blue/red, motors 1 and 4) and two counterclockwise (black/white, motors 2 and 3).

Strip, tin, and solder the battery cable to the flight controller (Figure I), carefully following the polarity markings: red wire (positive) to the VCC pad, black wire (ground) to GND. There isn't circuit protection at the battery input, so you will likely burn out the flight controller if you connect the battery backward.

3. ASSEMBLE THE COPTER

Install the control box assembly into the frame with two M1.2 screws and gently install the motors in their corresponding mounts.

Secure the flight controller and motor wires in place using a rubber band or hair elastic attached to the tabs located at the bottom of the control box (Figure J). Do not install the propellers at this time.

Secure the power cable in place with a cable tie so that it does not interfere with access to the USB port (Figure K).

4. BIND YOUR RECEIVER AND TRANSMITTER

Follow the manufacturer's binding procedure for your radio transmitter and bind it to your quadcopter. To do this, press and hold the receiver's Bind button (or ground out the Bind pad) and connect the

battery to the quad. The blue Bind light should illuminate and stay lit. You can now bind it to your radio.

Double-check that binding was successful by powering off your quadcopter, followed by your radio, then turn on your radio and then the quadcopter. The blue light on the receiver should start flashing, indicating a successful bind. Power down your radio and quadcopter and get your computer fired up!

5. CONFIGURE THE FLIGHT SOFTWARE

Cleanflight is free, open source flight controller software (github.com/cleanflight) for multirotors and fixed-wing planes. Here's how to set it up for your tiny quadcopter.

Using Google Chrome, visit the Chrome Web Store and install the Cleanflight Configurator. I installed version 2.1.0 at the time of this writing (Figure L).

Plug in the flight controller using a micro-USB cable and connect it to your computer. Launch the Cleanflight Configurator and select the Firmware tab (Figure M). Select the correct COM port at the top right corner of the window, set your board type to SPRACINGF3EVO, and choose the latest firmware.

Load the firmware either online or locally, if you downloaded it earlier, then press the Flash Firmware button. The firmware should now be loaded onto your flight controller.

If you get a "Failed to open COM port" error using Windows, try opening and closing the port using a secondary program like Arduino.

Press the Connect button and you'll be presented with the Setup screen (Figure N). Here you can verify that your flight controller is operating correctly by moving your quadcopter around and watching the visualization respond accordingly. Place your quadcopter on a level surface and press the Calibrate Accelerometer button. This process usually takes about 15 seconds. Do not touch the quadcopter while it calibrates. It will complete when the Pitch and Roll readings stabilize near 0°.

Select the Ports tab to configure your ports (Figure O). Make sure that both UART1 and USB VCP are set to 115200 baud and all other settings are disabled. Press the Save and Reboot button and reconnect your quadcopter if it doesn't do so automatically.

Select the Configuration tab and verify "Motor PWM speed Separated from PID speed" is enabled, "MOTOR_STOP" is disabled, and "Disarm motors regardless of throttle value" is enabled (Figure P on the following page).

Scroll down and verify that PPM RX Input is selected for Receiver Mode. Also check that Telemetry is enabled and Transponder and RSSI (Signal Strength) toggles are disabled under System Configuration settings. Press the Save and Reboot button

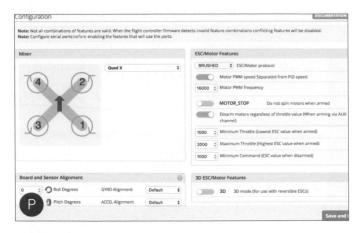

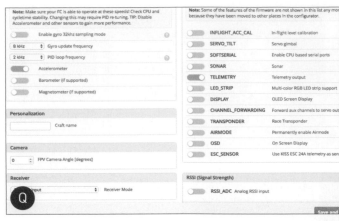

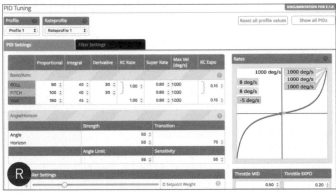

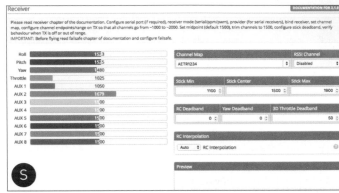

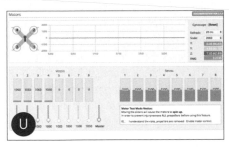

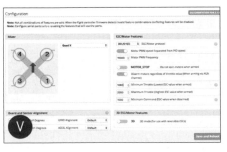

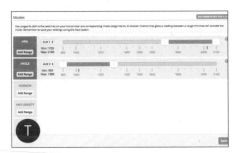

before moving on (Figure Q).

Skip ahead to the PID Tuning tab and press the "Reset all profile values" button (Figure R). This will use a pre-determined set of values that should serve as a good starting point and can be adjusted later to suit your flying style. (For other PID values, check the Thingiverse page for the KBWhoop or look at other users' values for quadcopters like the Tiny Whoop and flight controllers like the Beecore.) Click the Save button before moving on.

Select the Receiver tab and power on your radio transmitter. The blue light on your receiver should begin flashing and you should see bars populate the channel indicators (Figure S). If you move the stick on your radio, the bars should move as well. Double check that your channel directions and trims are set so that channels 1–4 read **1500** and the 3D quadcopter in the Preview graphic moves correctly. Save before moving on.

Select the Modes tab and select Add Range for both ARM and ANGLE modes. I configured my controller to use channels 5 (AUX 1) and 6 (AUX 2) for my arm and angle switches. Slide the ranges so that they correspond with your switches and you can test their function (Figure T). Save before

moving on.

Plug in your battery and select the Motors tab. Here you'll set the minimum value to start all of your motors (Figure U). To do this, toggle the Motor Test Mode switch and move the Master slider up until the motors begin to spin consistently — then record that throttle value.

Also verify that each motor is spinning in the correct direction and their labels are correct. Move each slider, from 1 to 4, individually, and cross-reference its location and rotation direction with the graphic.

Return to the Configuration tab and set your Minimum Throttle (Figure V) to the value you recorded; in my case the value was 1060. Now press Save and Reboot, and your flight controller is ready.

6. ATTACH THE CAMERA

Remove the micro-JST connector from the end of the camera's power wires and strip and tin the ends. Feed the wires down through the appropriate hole in the canopy and press the camera into place on the canopy. You may need to use a piece of double-sided tape to keep the camera from popping out during crashes.

Solder the power wires onto the large capacitor located near the flight controller's

USB jack (Figure W). Take care during this step as it is very easy to unintentionally short-circuit the electronics surrounding the capacitor.

Alternatively you can solder the camera's power wires onto the same pads that the battery cable uses.

Secure the camera assembly in place with two M1.4 screws (Figure X). The quadcopter is now ready for a pre-flight check!

7. PREFLIGHT CHECK

Attach the second hair elastic to the tabs on the bottom of the control box and secure the battery in place (Figure Y). Plug in the battery and set the quadcopter on a flat surface. Power on your radio and arm the quadcopter. The motors should begin to idle. Make sure that they are rotating in the correct direction and respond correctly to your throttle and attitude input.

Power on your FPV goggles and search for the channel the camera is on. You may need to adjust the channel to find one with the least interference. In my experience with this camera, I have to fiddle with the channels a bit before flying, as any interference greatly affects its range.

Disarm and unplug the battery on your quadcopter. Install the four propellers in the correct orientation by aligning them on the motor shafts and pressing straight down (Figure Z). My propellers were labeled with an "A" for counterclockwise and "B" for clockwise. Try not to wrench or pry the propellers on or off the motors, as you will likely bend the shafts.

Your quadcopter is now complete and ready for flight!

FLY IT!

Strap on your FPV goggles and take your 3D printed micro quadcopter for its maiden flight! I hope you enjoy your micro quad and get many hours of use, racing it around. I think the best part is being able to sit right outside your house and use the surrounding area as a racetrack without worrying about harming anyone or breaking expensive components. And when you do break this copter, you can print new parts for pennies.

Try experimenting with motors featuring different KV ratings, lighter cameras, two- and three-blade propellers, and different frame materials until it's "just right" for your flying style. Ping me on Twitter (@atomkemp) if you make one — I am happy to give you some tuning tips to get you flying right.

Don't forget to check the KempBros blog (kempbros.github.io) for updates to this project and other exciting adventures in the world of FPV. There's also a huge community flying micro-scale quads, so check with your local modelers association or websites like RCGroups (rcgroups.com/micro-multirotor-drones-984) and Reddit (reddit.com/r/Quadcopter) to get advice and fly with others.

As always, make sure you follow your local rules and regulations concerning flying models with FPV. Have fun and happy modeling! ⊘

Adam Kemp, Hep Svadja, FliteTest

makezine.com/61

More Interesting Flying Projects

FPV NIGHT FLYING
makezine.com/go/infrared-night-flying
Who wants to quit just because the sun goes down? Fly your FPV drone at night with an infrared upgrade.

HOVERSHIP: 3D-PRINTED RACING DRONE
makezine.com/projects/hovership-3d-printed-racing-drone
3D print this full-sized racing chassis and compete with the pros.

BUILD YOUR FIRST TRICOPTER
makezine.com/projects/build-your-first-tricopter
With smoother flight and better videos than quads, why not go tri?

PROTECT YOUR R/C ELECTRONICS FROM THE ELEMENTS
makezine.com/projects/how-to-protect-your-rc-electronics-from-the-elements
A good day of R/C fun can be ruined by even a relatively small puddle. Protect your electronics from a dunk with a few readily available products.

A

B

C

ADAM WOODWORTH is a lifelong aviation nut who's been flying R/C since childhood. He's currently a hardware engineer at X, Alphabet's innovation lab.

Doctoring Drones
Written by Adam Woodworth

Let your imagination take flight with toy-inspired quad builds

Three years ago I integrated a quad rotor into an old Hasbro *Star Wars* Speeder Bike, unknowingly setting myself down a path that resulted in millions of YouTube views, hundreds of new maker friends, and an insatiable desire to make things fly that simply shouldn't. Now whenever I walk down a toy aisle, I'm constantly on the lookout for my next build.

WEIGHT
When it comes to flying, weight is king. Using plastic toys is the easiest way to create a lifelike model, but they're heavy. A rotary tool is your friend. Use a sanding drum to remove any unnecessary material, and swap in lighter materials for non-structural parts. If you don't want to use a donor vehicle, or can't find one, there are thousands of free papercraft plans of most popular sci-fi ships. Recreating these with plastic or foam is an easy way to get a light, highly detailed model (Figure A).

LIFT
Your next consideration should be lift, and blending the propulsion system into the model as much as possible (Figure B). Black rotors are barely visible while spinning, and you can now find many rotor sizes molded in clear plastic, which look great on display. Multirotors have to tilt in the direction they want to go. If the rotors are installed level with the rest of the vehicle, this can make for a strange nose-down attitude in forward flight. Install the rotors at an angle so that the ship cruises around level.

DETAILS
Now it's time to decorate. Weathering or dirtying up is the easiest way to make it look realistic, and a simple paint wash can add depth and bring out details like panel lines. Add your own touches by scoring the part with a razor blade, then going back to work in the paint wash. Little defects add character and can cover up mistakes. When possible, lay your parts flat for painting (Figure C).

MATERIALS
Your creations might have some bizarre handling qualities, so getting them to fly well is an exercise in trial and error. These frequent encounters with terra firma can take their toll on light/fragile structures. To address this, I've recently shifted from using rigid foams like EPS and Depron to more compliant materials like expanded polypropylene (EPP). EPP can be sourced in a variety of densities, in sheet thicknesses from 2mm to 10mm. Thinner foams can be used to replicate most papercrafting techniques, and hold up well to repeated abuse. As an added bonus, EPP is solvent resistant, so it can be painted with pretty much anything and easily glued with cyanoacrylate (super glue).

Now get out there and get building! We live in a new golden age of flight. ◐

DIY Drone Recovery Parachute

Written by Andrew Morgan with Dr. Richard Chapman and Dr. Saad Biaz

Sensors determine if your aircraft's in trouble and bring it down safely

We have all seen a quadcopter in the sky, flying so majestically in one location — until it drops like a rock. Unlike fixed wing aircraft, quadcopters lose lift when the battery is depleted, or even when the craft is upset beyond its ability to recover.

We designed and constructed a ballistic parachute recovery system for small unmanned aircraft. Based on an Arduino microcontroller, the recovery system uses sensors to determine GPS coordinates, remaining battery voltage, and acceleration. If the system determines that the drone's battery is depleted, or that it is operating outside of prescribed GPS boundaries, or that the unit is in free-fall, the recovery system cuts power to the motors and deploys the parachute, lowering the aircraft to the ground at a safe velocity. These are the components. Find the complete build online at makezine.com/go/drone-recovery-parachute. ⊘

TIME REQUIRED:
6–7 Hours

COST:
$55

ANDREW MORGAN is a Ph.D. candidate at Yale University specializing in Robotics and interested in autonomous systems. This project was completed in the summer of 2016 at Auburn University under a Research Experience for Undergraduates program funded by the National Science Foundation. **DR. SAAD BIAZ**, Principal Investigator, **DR. RICHARD CHAPMAN** co-Principal Investigator.

Ⓐ The recovery system is controlled independently of the drone's flight computer through the use of an Arduino Nano **microcontroller** powered by a separate 7.4V LiPo battery, to ensure proper operation of the recovery system in the event of a drained main battery. Each hardware component is connected to the microcontroller via the digital or analog I/O pins.

Ⓑ Acceleration components in x, y, and z directions are read according to voltage values from the **accelerometer** module.

Ⓒ The **GPS** module is powered through the dedicated recovery system battery and transmits NMEA data to the Arduino via a RS232 serial connection.

Ⓓ The **5V relay module** cuts the power to the drone's motors via an internal switch.

Ⓔ The **voltage sensor** unit acts as a 4:1 voltage divider, providing a voltage range within the limits of the Arduino's analog-to-digital conversion circuitry.

Ⓕ To save battery and Arduino processing power, the **servo motor** that deploys the parachute is set to close initially and then virtually detach from the system.

Ⓖ A **parachute** is nothing more than a piece of material (nylon works great) with some string to tie it all together — look at designs and make your own. The parachute can be constructed of a PVC tube, large spring, baseplate, 3D printed door and servo motor holder, and a cheap servo motor. The servo motor merely holds the door back until deployment, when the parachute fabric is launched outward by the internal spring. Note: We used an older version of the Mars Mini in our original build, but premade chutes can significantly up your cost. ⊘

Hep Svadja, Andrew Morgan

Cyborg AND THE SINO:BIT

Written by Naomi Wu

Shenzhen's prolific maker details her journey, her inspirations, and putting together China's first certified Open Source Hardware project

I'm **Naomi Wu** (@realsexycyborg on Twitter), a 23-year-old maker and hardware enthusiast from China. I live in Shenzhen, also known as the "Silicon Valley of Hardware." Chances are your phone or computer was made here — maybe even by a girl I grew up with. It's a city straight out of cyberpunk, on the cutting edge of tech, and it's growing at an amazing rate. I'm right in the middle of it.

Shenzhen is located in Guangdong province, Canton by the old spelling, so I'm Cantonese. It's right across the border from Hong Kong and, at less than 40 years old, is one of the youngest cities in China. Our culture reflects that; I'm told we're a bit like New York — people come from all over to change their life and make their fortune. We are fast moving and ambitious, but tend to be less conservative than other Chinese cities. "Local" cuisine is from all over China, because locals are from all over China — and increasingly the West.

Shenzhen used to be known as the Shanzhai capital — Shanzhai being our term for copycat or cloned products. Obviously, we know this business model of just taking other people's ideas and making them is not sustainable. We'd all be factory workers forever if we did that. So there's been a tremendous amount of thought and resources put into fostering an environment of creativity and innovation. More than anything, I'm the result of this supportive environment, millions of people sharing a common goal and value — to have, and produce the products of our own ideas. To be creators, not laborers. So when I show off one of my creations in public, be it my wearable 3D printer or skirt made of infinity mirrors, everyone is delighted, because as foolish as they are, it's creative foolishness and something we are all striving for. Seeing some rich boss pull up in a fancy car, this does not make anyone happy. But taking a selfie with the weird local girl riding the metro with a working 3D printer on her back? Everyone smiles because this represents the creative, interesting city we all want. It is also something that does not come easily given many of our traditions.

Photos Courtesy of Naomi Wu

Hikaru Skirt

Pi Palette

Poor Girl's Monocular Display

Infinity Skirt

LED Heels

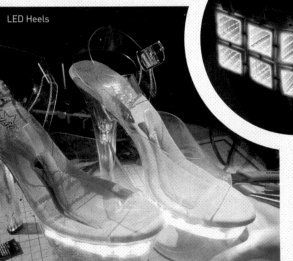

When I show off one of my creations in public, everyone is delighted, because as foolish as they are, it's creative foolishness and something we are all striving for.

A CRASH COURSE IN MAKING

I was educated locally here in China — what in the West you would call public schools. I come from a fairly typical working-class background. I think the education I had was good given the resources available — if not as comprehensive as a Western one. It was, however, strong on math and science which I am grateful for now given my involvement in tech. I was a bit of an English geek, studied constantly and watched English TV late into the night, every night since I was a child. Because of this, I was able to win awards in local contests for my English. This proved to be an advantage when I attended college as an English major and needed to earn some extra pocket money, as it enabled me to learn to code online and become a freelance web developer. Web development exposed me to startup culture, then to the local Shenzhen hardware startup and development scene. There are many Westerners in Shenzhen doing hardware development, and from talking to them I heard about maker culture, hardware hacking, and infosec — enough to spur my interest to study these things more on my own and eventually become a maker.

My first hardware project was for a Maker Faire after-party in 2015. It was a skirt under-lit

Pentesting shoes

At work high above Shenzhen

Bangkok Mini Maker Faire

with LEDs based on a Japanese design by Kiyoyuki Amano — the Hikaru skirt. I had access to a 3D printer, had completed the Tinkercad tutorials, and had been using them to make some small gifts for friends. A simple enclosure for the control board and battery I had purchased was just a matter of printing and revising the design a few dozen times until everything fit. From my software development and use of open source tools I knew the importance of attribution so was very careful not to take credit for the skirt's concept. The creator and the Japanese maker community appreciated this, because it's unusual for Chinese to be so careful about attribution, and they reposted the pictures of the skirt. This was an important early lesson for me — you get more respect through proper attribution than you do by claiming others ideas as your own. Eventually, the skirt made its way onto Western websites. For a regular girl from a city of 12 million, to suddenly have her picture in the West — this was quite something. It's easy to feel invisible and lost in a mass of humanity here. I'm not ashamed to admit that more than a little bit of vanity drove those early days of late nights over the soldering iron. But I matured and while I still love the

presentation and performance of showing off something I make, love of the build itself has surpassed these elements. In the following two years, I kept up as good a pace as I could while working full-time, with a new project every two months or so. I created 3D printed heels with pentesting (hacking) tools built in, arm-mounted micro drones, a skirt made of infinity mirrors, a burlesque-inspired top made of LCD shutters (I had something underneath for modesty, of course), a makeup palette with a Raspberry Pi built

had not wanted to be the sort of person who stood out in a crowd it's unlikely I would have been attracted to making and fashion tech — activities that are ideally suited to this kind of creative personal expression.

The Asian tradition of pretty girls doing creative things, I suppose Westerners would be most familiar with the Japanese geisha, originated in China, of course. We've been doing it for thousands of years and there is no conflict; it is our ideal. People from other parts of the world, with different tastes and artistic

of the Westerners coming to Shenzhen to have hardware made were men, and most of my online friends in maker education were women. These women knew exactly what was needed in their classrooms — Pi Hats and Arduino Shields, specialty RGB LED displays. All sorts of things, all very specific, and very practical ideas. But they had no experience in manufacturing and were a bit reluctant to get on a plane to a non-English speaking country and make a go of it alone. For that matter they were a bit concerned if they could even get it made after crowdfunding it. There are just not many women they could speak to who had produced simple hardware in China successfully on their own.

This coincided with an interest I had in the BBC micro:bit. The micro:bit is a brilliant piece of engineering for education, but like many open source projects part of what makes it so great is the ease with which it can be tailored to different needs. From talking to maker education people here in China the first, most basic step of programming — Hello World — was being glossed over or bypassed entirely. "Hello World!" just does not have the same thrill when it's in a language you

Making is not a purely technical pursuit, there is a bit of performance art to a lot of it as well.

in for more network pentesting, a device for a small drone to deposit a Wi-Fi-hacking payload and fly away, and more.

Of course, since I became an adult I have had a very flamboyant personal style — I know that played a big part in the exposure I received for these projects as well. Making is not a purely technical pursuit, there is a bit of performance art to a lot of it as well. Presentation often plays a role and you can't unscramble an egg and separate that from the technical merits of the project. But if I

traditions may find it all a bit surprising. I do what I can to accommodate but I am Chinese after all, so mostly concern myself with our traditions. So long as old ladies laugh, smile, and take pictures with me I don't worry too much.

LEVERAGING SHENZHEN WITH THE SINO:BIT

In addition to the projects I build myself I recently collaborated in the creation of an educational board called the sino:bit. The project came about as I began to see a pattern, most

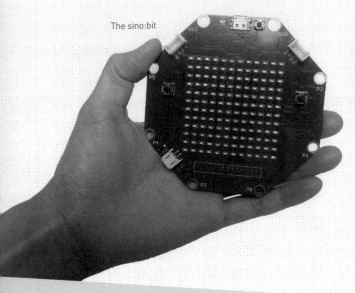
The sino:bit

Wearable 3D printer

Young Naomi

can barely read. U.K. children would not have much interest in spending an hour learning to write "你好，全世界！" on a display or showing it to their uncomprehending parents. So instead Chinese kids wanted to make a robot. The micro:bits were almost never being used here without an add-on board for driving motors. Troubleshooting the interface between the micro:bit and the connector slots on those boards was taking up valuable classroom time.

I came across the German version of the micro:bit — the Calliope Mini. It seemed that many of the issues we had in China — localization and an easy way to drive motors — the Calliope team had already solved. So with their permission, I proposed a Chinese derivative of the Calliope Mini — largely the same, backward compatible with both the Calliope and the micro:bit but with one important difference. The sino:bit would have a 12×12 LED matrix instead of 5×5 so that it could display any non-Latin language — not just Chinese characters,

but Japanese, Arabic, Thai, Hindi. Kids could experience that first thrill of "Hello World!" in their mother tongue. Not only that, their parents could read messages — holiday greetings, "I love you Mom," even religious messages programmed by their children. It is hard to overstate the role of the written word in Chinese culture and education, suffice to say it is inseparable. The sino:bit takes the amazing engineering that went into the micro:bit and Calliope Mini, and tailors it to fit Chinese culture and educational tradition — instead of expecting us to change to fit another culture's language and traditions.

So with this project and the concerns about manufacturing difficulties faced by my female friends in maker education in mind, I contacted a local electronics manufacturing company called Elecrow. The idea was they would do the

> **"Hello World!" just does not have the same thrill when it's in a language you can barely read.**

engineering on the sino:bit to bring the idea to life, and even though we are all Chinese we would conduct and document the entire process in English. I would even visit the factory speaking only English to the taxi driver, showing the address written on a piece of paper in Chinese.

All this to answer one question: Could a typical online friend of mine describe her idea and have it manufactured, entirely in English? Either simply over email or with a visit to Shenzhen to oversee delivery? I envisioned a Western woman in maker education, capable of programming and teaching Arduino or other basic electronics, but without PCB layout or electrical engineering experience. What sort of pitfalls were there to be aware of? Can you really just have someone else do all the engineering based on your general idea of

how it should work? I made certain it was challenging for Elecrow — I refused to reply to Chinese messages when they wanted something clarified, was deliberately vague, constantly told them to do what they think is best, and in general insisted on the sort of initiative that I knew Westerners encouraged but that we find difficult. I often joked I was "red teaming" Chinese hardware development — taking a joke from infosec.

It was my first hardware development project, and a learning experience for everyone. But in the end, we became the creators of the first Open Source Hardware Association certified project in China. The publicity has more than paid for the time they put in — something that other companies have noticed. Those companies have started to reach out to me about how to go about getting involved in open source projects of their own, to raise their profile in the hardware community and show they understand these values. All this along with the growing interest in the sino:bit

sino:bit production line

Naomi at Elecrow developing sino:bit

Becky Button

as an educational tool for Chinese children, is, of course, immensely gratifying.

CHINA'S OPEN SOURCE CHALLENGE

Recently, in order to have more content on my YouTube channel in between builds, I've started shooting videos reviewing local products made here in Shenzhen. I'm quite careful not to endorse anything shoddy, or poorly made, and this has lead to local companies reaching out to me asking me to help make Western customers more aware of them. This creates some interesting opportunities for advocacy and cooperation between Chinese and Western hardware communities.

My open source interests and video creation recently converged with one specific interaction. There's a fairly well-known 3D printing company here in Shenzhen — Creality 3D. They had invited me to shoot a video of their factory. It was a little ways out of my way — about an hour from my house, so I had been dragging my feet a little. Around the same time the

Marlin firmware development team — that's the program that runs a lot of 3D printers — was complaining that Creality had not been publishing the changes they made to the code as dictated by the license terms. So I thought to myself to have a try and half-jokingly told Creality I'd visit the factory if they published the code. They immediately agreed. This was quite a surprise because open source compliance is normally quite tough in China. So I went and did the best job I could with the video, in part because it's a pretty typical Shenzhen factory and I think it's important for people to see most are perfectly reasonable places, and that the bad factories are few in number and quickly disappearing. But also because I wanted to do everything I could to support and promote a Chinese company willing to at least take a very unusual first step towards abiding by open source community standards.

I did the tour, spoke to the workers and bosses, and on my way out they very kindly offered me one of their 3D

printers — a very large one called the CR-10s. Now, I'm already running out of room in my little apartment workshop. I have one large printer and an additional one would just be wasteful. So I was going to refuse but had a thought, so said — "Boss, everyone has seen your printer reviews, but you have no community engagement. Why not do what Josef Prusa and LulzBot do and show not just your product, but the values of the company behind it? I know a teenage maker in America — Becky Button. She's sharp as a tack, makes great things, but does not have her own 3D printer or very easy access to one." I took out my phone and showed the boss Becky's project that I had given her a bit of advice on — sandals that can disconnect Wi-Fi around them. Well that was that — we are Chinese after all and if a smart kid needs something educational, and there is face to be had by providing it, it's going to happen. They loved the idea and Becky got a printer. Culturally for us this is quite a big step and a

success I hope to be able to repeat with other companies. Tools and making are closely associated with men in the mind of many manufacturers, so it can be hard for female makers, educators, and vloggers to get sponsors and review units. I'd really like if I could help address this.

I strive to facilitate this sort of interaction, to talk to Chinese companies and say, "This is why people are annoyed with you," or talk to the maker and hardware community and say "OK this is why this is happening at this company." It's very exciting and rewarding being able to act as a bridge, in even the smallest ways because there is a tremendous amount of miscommunication but honestly far more similarities in terms of creative spirit and values than there are differences between China and the West. I'll always want to make things, but this sort of combination of diplomacy and evangelism for the hardware and maker communities is really fulfilling and something I hope to be able to do more of. ⬤

Photos Courtesy of Naomi Wu

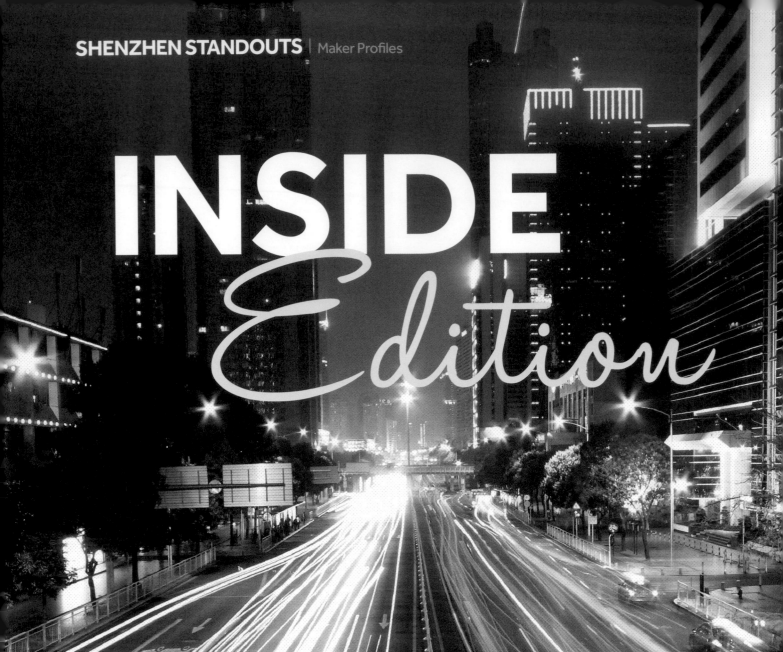

INSIDE
Edition

Shenzhen's locals leverage the region's assets to advance its global maker leadership

The center of the consumer-electronics universe, Shenzhen and its environs are a business-first mix of factories and parts-suppliers, designed to quickly manufacture the devices that the world craves. As it has grown into one of China's major cities in just a few short years, the region's population — those assembling all your gadgets — has started to express its maker spirit, with notable individuals, hackerspaces, maker ed programs, and even a Fab Academy taking prominence on the global stage.

We asked some of the local participants who are helping establish the maker scene to share their stories.

Jess Yu / Adobe Stock

Vicky Xie

Courtesy of Vicky Xie

I am the director of International Collaboration of the Shenzhen Open Innovation Lab. I came into making in the most unexpected way in the past three years, helping set up Fablab Shenzhen as my first job after college, while seeing how making has transformed the image of my native city of Shenzhen.

I majored in English and after school I got a job in the international department of Shenzhen Industrial Design Association (SIDA) and was assigned to work on the upcoming first Shenzhen Maker Week and the gathering of 10 Fab Labs in June 2015. I had no idea about makers and the Fab Lab but had to learn about them in a short time. In the Shenzhen Maker Week event, my team organized exhibitions of Fab Labs including Boston, Barcelona, Taiwan, and Japan, and hosted presentations by pioneers like Neil Gershenfeld, Tom Igoe, Lyn Jeffery, and Tomas Diaz. A couple months later, I was part of the delegates to FAB11 in Boston that took on hosting of FAB12 in Shenzhen in 2016. I decided to learn more about making and became the director of International Collaboration of the newly founded Shenzhen Open Innovation Lab (SZOIL) to facilitate the collaboration between this city and the maker community.

I also decided to sign up for the Fab Academy, starting in December 2015. There, I learned how to envision, prototype, and document my ideas through many hours of hands-on experience with digital fabrication tools. I started to learn 2D and 3D design, electronics design and production, producing circuit boards, using a variety of sensors and output devices, programming AVR microcontrollers, and other interesting experiences such as molding and casting. There were about 10 people signed up for the first Fab Academy offering here and I was one of the five who graduated.

Right after FAB12, I joined the core organizing team of the 800-square-meter Maker Workshop in the National Mass Innovation and Entrepreneurship Week — a major national event to promote innovation policy by the Premier Li Keqiang. We put together a full Fab Lab in the event space and prepared workshops for thousands of students and visitors over five days.

In 2016, we at SZOIL kicked off the Hello Shenzhen program with the Shenzhen Foundation for International Exchange and Communication as well as the British Council, for an in-depth exchange between U.K. and Shenzhen makers. Ten makers based here have been selected to take part in three week-long residencies with some of the U.K.'s leading makerspaces and creative organizations including Access Space, Central Research Laboratory, FACT, Impact Hub Westminster, Lighthouse, Machines Room, and Makerversity.

In the meantime, SZOIL is also increasing its engagement with the Fab Lab and makerspaces in the emerging markets. I organized the Belt and Road Maker Summits of International Maker Cooperation in 2016 and 2017 bringing the makers we have been working with in Nigeria, Ethiopia, Peru, Pakistan as well as Shenzhen partners.

We would like to build up a platform for better understanding of the ecosystem for collaboration between this city and the countries and areas alongside the "Belt and Road" as well as seek more potential opportunities among startups, makerspaces, and accelerators. The latest project is working with Impact Hub Accra to help them produce a low-cost biodigester to turn scraps into fertilizer and gas.

I am grateful to have the opportunity to get involved with making in the past three years and be in the position to collaborate with all the amazing makers and my native city of Shenzhen.

Courtesy of SIDA, Vicky Xie

Shirley Feng

I am the secretary general of Shenzhen Industrial Design Association. I came to Shenzhen in the 1990s and started my career here as a computer programmer. I guess this makes me one of the original makers in Shenzhen, as I used to go to Huaqiangbei to put together computers and other equipment for work.

In the past two decades, the city of Shenzhen has undergone drastic changes, growing into a global manufacturing hub and then into the global hub of design and innovation. I have had the fortune to witness these amazing transformations and participated in some of them.

Realizing that Shenzhen cannot grow forever by simply producing things, I became the general secretary of Shenzhen Industrial Design Association (SIDA) in 2008. At that time, the government and the business in Shenzhen focused solely on GDP and revenue growth; industrial design was considered lightweight and unimportant.

Over the years, my team at SIDA and I have worked hard to promote the importance of industrial design to government and business. We have grown SIDA from 49 institutional members to over 700 in the past 6 years, which makes SIDA one of the largest industrial design association in the world. SIDA's members, along with over 100,000 working industrial designers in Shenzhen, have helped the Shenzhen manufacturers continue their growth by moving up the value chain from OEM to ODM in the past decade.

Today, Shenzhen has one of the best government policies in the world to encourage creativity and innovation in industrial design, and the Shenzhen Industrial Design Faire 2014 has became the largest industrial design event in the world.

This is a young city under constant change. But the change is far from over. With SIDA, we have brought better design here. And an eager young city like Shenzhen is always ready to learn more and grow better. In 2012, the maker movement started to get traction and Shenzhen hosted the first Maker Faire in China. And by word of mouth and a lot of blog posts, Shenzhen quickly become the maker's favorite city, especially Huaqiangbei with its blocks of giant malls of electronics.

In 2014, I had the pleasure to visit FAB10 in Barcelona and was very impressed with the passion, creativity, and innovation exhibited. There and then, I knew this was what Shenzhen needed: passion, creativity, and innovation! I started to work with my team in SIDA on how we can promote the makers and Fab Lab here. Working with the government, on June 18 of 2015, we invited 10 Fab Labs around the world to join us at the Shenzhen Maker Week and I led the team to FAB11 in August of 2015 to pledge Shenzhen as the host of FAB12 in 2016. In August of 2016, I had the pleasure to welcome over 1,000 makers from over 100 Fab Labs in 78 countries gathering in our city.

When the National Mass Innovation Week took place here this past October, I had the honor to present the progress of Shenzhen's leading the nation in the maker movement to Premier Li Keqiang during his stop at our Maker Workshop.

The maker movement always will be part of the Shenzhen DNA. As an early arriver in Shenzhen, I have benefited from the ecosystem in the city in my career going from a programmer putting together machines needed for my work in HQB, to the secretary general of SIDA to help and promote the Shenzhen industrial designers to help Shenzhen manufacturers move up the value chain, and to now building the bridge between global makers and the Shenzhen ecosystem to help more people realize their ideas and dreams. I am grateful to have grown with Shenzhen, the city of makers.

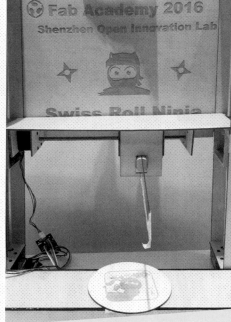

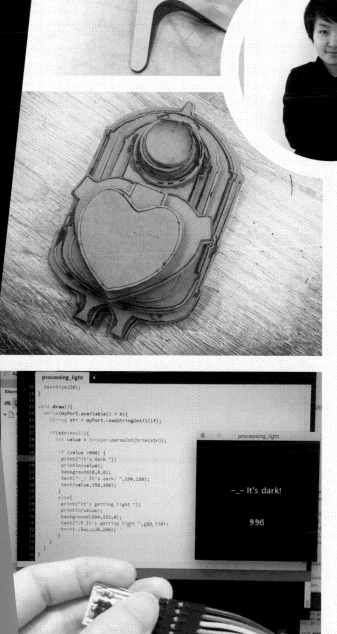

Lin Jie, aka OO

I might call myself as non-mainstream, wandering around, designer-who-is-keen-to-make-tangible-things type maker. As an interaction designer, I've been interested in building interactive 「smart」 gadgets. I bought my first Arduino board in 2012, and tried to learn some basic knowledge of making digital things. But not until I enrolled at Fab Academy in 2016 did I begin to treat myself as a future maker. After finishing all the courses and projects, I finally have some confidence to refer to myself as a junior maker.

My favorite project I've made is a butterfly music stand I did during the Fab Academy program. The assignment was to make a large model using the CNC machine. I had been considering buying a music stand for my guitar practice, so I designed the style and made a demo first, then used a CNC machine to cut out all the pieces and assembled them.

SOME OTHER FAVORITES:
- **My final project — a mini soundbot:** If you wave or move hands in front of the soundbot, it will convert the distance between your hand and the sensor into music notes. You can play a song with your hand but without touching anything. It's like a geeky, modern theremin.
- **Swiss Roll Ninja:** An automatic roll cutter made by our study group in Shenzhen Open Innovation Lab.
- **Multilayer Minion:** A minion from the animated movie, made of stacked laser-cut pieces.
- **Light detector:** I made a PCB with some light sensors to tell if it's dark in the room.

My maker advice for others: Don't be hesitant, as I used to be. Make more and you will have more fun. Makers never get bored, right?

Find me online at uegeek.com.

Lin Jie

Carrie Leung

I am a San Francisco native who has been living in Shenzhen for the last six years. During the day I am the director of maker ed at Shenzhen American International School. We are a 100% project-based learning school with maker embedded classes. What this means is our school's makerspace is more than just an extension of a one-hour-a-week STEAM class, an elective class, or an after-school club. The use of our makerspace is a learning method for every student's full-time curriculum. It is referenced as often as a book, a teacher, or the internet. Our school is driven by class projects and community problems that the kids work on. Our students embody the maker mindset and the making is part of their full time education.

When I'm not at school, I devote my time bridging, enabling, and empowering the communities in Shenzhen through making, sharing, and collaboration. I have a nonprofit makerspace, SteamHead, which is the home base for many of these initiatives. I strive to create free, bilingual, easily accessible, and open source platforms. There are four initiatives that are very important to me:

● Giving young makers a platform to express themselves
● Bridging the public, private, and international educational communities in Shenzhen
● Enabling classrooms to use maker skills and project-based learning methods
● Empowering girls and women by building confidence through experience and shared knowledge.

We have just started the Girls Can campaign, which is a program that creates and sponsors women- and girl-led workshops and events, and also documents and shares what individual women and girls are already doing.

Making has always been part of who I am. I've always had that kindle of curiosity or imagination that flames into something tangible. I remember when I was four years old and I took my father's record player apart. By the time anyone got a whiff of what I was doing, it was too late. I had pieces everywhere! To my disappointment, I did not unwrap the mysteries of how the turntable produced melodies, but I did make one incredibly angry father.

A recent favorite project was going from silly concept to awesome wearable in 20 minutes. Shannon Hoover of MakeFashion was in town and he showed me how to use his StitchKit board. I wanted to be a dinosaur, so board, hot glue, chopsticks, cloth scrap, and some minutes later... BAM! I was dino-glowing it up. It never ceases to amaze me how tech today allows us to go from fiction to reality in a matter of moments.

Nowadays, when I make for myself and my friends, I make for fun and laughs. When I make for the community, I make with positivity and openness. When I make for the future, I make with purpose.

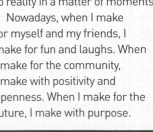

SZOIL and SIDA

Lit Liao

I run Litchee Lab as a maker ed education team and also as a makerspace. Six years ago, I joined Seeed Studio, an open source hardware company in Shenzhen, which brought Maker Faire into China. I was responsible for education products, so I'm involved in a lot of workshops designed and run in Chaihuo makerspace, the first makerspace in Shenzhen, and had lots of online communications with makers all over the world to understand user experience when they use Seeed's products. Some things in maker culture really touched me, such as its open source ideals, and encouraging everyone to step out of their own comfortable area and try something new. With online guides and resources, I found myself growing confidence with my skills about making.

And then three years ago, I quit my product design job and started Litchee Lab to explore a sustainable way for a makerspace to survive and find the answer to how maker ed can grow in the Chinese community. Litchee Lab is a grassroots Fab Lab in Shenzhen that supports itself with only membership fees and education services, no big company or government funding. We provide an as-free-as-possible platform for teenage and adult makers to explore their own ideas.

The operation team of Litchee is devoted to education. We are the first team in Shenzhen to have included maker ed curriculum development into our mission. And so far, we are cooperating with 20 schools in Shenzhen to build their own maker ed curriculum. The mission of the education team is helping local kids develop their creativity by providing a platform we designed by combining international resources and local insights.

For three years, Litchee Lab accepted members from eight countries all over the world. Though we are a startup makerspace, it is one of the most recommended makerspaces in Shenzhen's foreigner maker community due to openness and freedom (24/7 door access, super convenient location, good community atmosphere). Litchee Lab is involved in a China-U.K. maker exchange program organized by the British Council since 2015. And we just registered as an official makerspace platform in SZSTI library so that members can apply for Shenzhen government funding through us. ◗

Lit Liao, Byron Wang, Jin Dongjun

SHENZHEN
Goes Pro

Written by Violet Su

Alongside the exhibits, workshops, talks, and performances at Maker Faire Shenzhen this November were some new elements worth celebrating and sharing.

4 MAJOR DIFFERENCES

1. THEME
Unlike previous years, we adopted a theme for this year's Maker Faire — Maker Pro. As Shenzhen is known as the hardware capital of the world, there have always been more hardware startups than at other Faires, so we decided to talk about what we define as maker pro. We look at maker pros as those who might eventually carry forward as a startup or a skilled professional. They come in many different forms and backgrounds and have experienced different journeys in life. Some start as hobbyists, some as professionals, others as communities, while there are those who were entrepreneurs from day one. But they share one thing in common — a passion for making their ideas a reality, and sooner or later, taking them to the next level to face the market.

2. VENUE
We wanted this year's Maker Faire to take place at a location with a strong connection to maker pros, and we found a

Maker Faire returns to the city with a new venue and theme

great partner — Shenzhen Polytechnic. This is one of the top technical institutions in China, with a specific emphasis on training in a wide range of subjects and skills from mechanical engineering, to automobiles, to design and visual arts. And with a maker center (soon to be registered as a Fab Lab) in the school, Shenzhen Polytechnic encourages students and teachers alike to gain hands-on experience, and to go from theory to practice and eventually to business. Organizing the Faire in an institution of higher education also conveys how the maker culture is strongly connected with education and is an outlet for makers to go pro in this city.

3. MAKER PRO EXHIBITION
Along with the projects that applied to be part of the Faire, organizer Chaihuo, which also manages maker pro-focused makerspace Chaihuo x.factory, curated a Maker Pro Exhibition that highlighted 11 maker pros and their products

and growth path, providing inspiration to others about how to take their projects to another level. Through this they told their tales of how they have evolved as makers to maker pros and hope to inspire many more people to begin their own journey.

4. MAKER ED SESSION
We held the first-ever Maker Education session at this year's Maker Forum, co-curated by local educators Carrie Leung, Joseph Strzempka, and James Simpson. We invited speakers from different education communities (schools, makerspaces, parents, government, or other organizations) to discuss the ways in which innovative learning has caught on in their own respective communities, and how these communities have come together to share and collaborate, encapsulating the spirit of making.

Apart from these exciting new firsts, here are some highlights of this year's Faire.

MAKER EXHIBITS
This year we gathered 156 maker teams from all over the world to show their projects. Among the makers, 65% were Chinese, and 35% were from overseas. The projects can be divided into seven categories, with software and hardware projects taking the highest percentage of 41%, followed by education projects of 25%, and an increasing percentage (15%) of interactive arts projects.

A sampling:

- **IRON MAKEY ROBOT** made from recycled beer bottles and metal parts by mosaic artists Ziyao Lin and Cynthia Shi with a group of local volunteers
- **DREAM CATCHER** by Taiwanese artist Ty Chen
- **PIXELMAN** by Ty Chen
- **WATERLIGHT GRAFFITI** created by French artist Antonin Fourneau
- **WOLVERINE CLAW AND ARMED TRACKED VEHICLE** hosted by Tangtang Cai, a maker from Wuxi, China

Chaihuo x.factory Maker Pro Exhibition

Iron Makey Robot

Dream Catcher

Waterlight Graffiti

- **HUGE EIGHT-LEG MOUNT** by Ziping Chen, a maker from Jiangxi, China
- **BLOOMING** by Alt+T team from Guangzhou, China
- **SIMPLE ANIMALS** by Eunny from Korea, who traveled to many Maker Faires around the world in 2017
- **LASER-CUT AND LAMINATED ART WORKS** by Ketian Zhang, a maker/architect from Tianjin, China
- **UPCYCLED EDUCATIONAL AND INTERACTIVE PROJECTS** by the Tomonic team who brought 50+ projects from Thailand and took 2 days at Chaihuo x.factory to build the music fountain, a main attraction at Maker Faire Shenzhen this year.

WORKSHOPS

There were 27 different workshops this year — more than 50 if you take recurrent workshops into account. These included virtual reality, PCB making, DIY 3D printer, project-based learning, woodwork, cardboard projects, escape game jam, music, laser cutting, sewing and fashion tech, upcycling and car making and racing, etc. And the workshop instructors were not only adults, but also kids. It's the workshops at Maker Faires that provide the most direct, visual/tactile, and hands-on experience to participants. You can find adults and kids, boys and girls immersed in the fun of making within the workshops. Such a not-to-miss part of Maker Faire!

NERDY DERBY

Nerdy Derby is one of the most popular projects at Maker Faire Shenzhen. With the license from Nerdy Derby Inc. and the joint efforts of Yan Chai Hospital, Lan Chi Pat Memorial Secondary School, Shenzhen's Oriental

Huge Eight-Leg Mount

Simple Animals

Workshops

Drone Experience Combat

Mario the Maker Magician

English College, Podconn Limited, and Chaihuo x.factory, we managed to build the tracks and add electronic sensors to the track to monitor the speed in less than two months.

DRONE EXPERIENCE COMBAT

This was the second time we hosted a drone combat exhibit with D1, and this year it was a bit different, with a focus on the experience. Instead of recruiting professional pilots in advance to join the combat, we encouraged everyone to join. There was a specific section where participants could learn how to control the drones from professional pilot mentors. After the quick training, participants joined the combat to have fun and to win prizes at the same time. More than 500 participants joined the Drone Experience Combat over the three days of the Faire.

Upcycled Projects

MARIO THE MAKER MAGICIAN

We were excited to have Mario the Maker Magician join us this year. In addition to their own booth in the maker exhibition area, Mario and manager Katie Marchese also staged six performances to all participants in the 3-day event. Kids and adults alike were amazed by the fabulous magic performance, which also perfectly illustrated that language isn't a barrier to maker culture.

XTALK

This was the second year we partnered with XTALK, a stage for lightning speeches. We had 24 speakers share their maker stories and projects, with 6 kids, including Wenjun

Shenzhen Polytechnic, Eunny, Team Chaihuo, MG Space, D1, Paola Paulino, Sherry Huss

Sherry Huss, vice president of *Make:*

MFSZ crew and volunteers

Women Maker Workshop

He, a 5th-grade local student, and Lauren and Ashley, the two youngest designers from the MakeFashion community in Calgary, Alberta.

MAKER FORUM

This year's Maker Forum consisted of four sessions, including "Visions into the Future," "Making in Industries," "Platform Builders," and "Maker Education." We invited 26 speakers to share their insights/stories/experiences at this stage. Out of the 26 speakers, 11 were women, including 8 Chinese women.

More than 2,000 attendees came to listen to the talks, 27% participated in the "Visions into the Future" session, 26% in the "Making in Industries" session, 23% in the "Platform Builders" session, and 24% in the "Maker Education" session.

MAKER PARTY

In the evening of November 10th, we hosted a maker party at Chaihuo x.factory, gathering makers/exhibitors/speakers and partners.

WOMEN MAKER WORKSHOP/ DISCUSSION

We also held a women maker meetup on the final day of the Faire. This meetup gathered 14 amazing women, including makers, educators, and community managers in the maker community, to share their experience, ideas, and suggestions for building a better ecosystem for encouraging women and girls to participate in making.

3-DAY HIGH TOUR

After the Faire wrapped, we hosted a 3-day High Tour for makers to explore this maker city. The (very intense!) schedule for the tour was as follows:

- **DAY 1:** Seeed Agile Manufacturing Center + PCB factory + Molding factory
- **DAY 2:** Tencent + XIVO design + BGI
- **DAY 3:** Chaihuo x.factory + Lab 0 + HAX + Huaqiangbei Electronic Market

32 makers/researchers/journalists from 11 countries joined the tour to get a closer look at this city known for hardware and manufacturing, and to learn about the resources that Shenzhen can provide — especially the techniques available for makers to adopt into their future designs and production. The 3-day tour was not only informative and fun, but also helped to build lasting friendships among participants.

Maker Faire Shenzhen provides us with opportunities to gather with friends old and new. It's a platform for us to connect with resources local and from afar. It's a carnival for us to celebrate making with kids and adults alike. With the conclusion of Maker Faire Shenzhen, we are grateful and would like to say "谢谢 (Xie xie = THANK YOU)" to all our partners, makers, friends, volunteers, and sponsors for your gracious help and support.

We look forward to meeting you all at Maker Faire Shenzhen 2018 and at other maker events around the world! ❷

See more photos and videos from Maker Faire Shenzhen at makezine.com/go/mfsz-2017

Written by Nicole Catrett

Rainbow Lightbox

Build a color quantizer out of mylar and scotch tape

NICOLE CATRETT
is an artist and co-founder of the Wonderful Idea Company, a creative design studio developing resources for the exploration of art, science, and technology through making and tinkering. You can find more wonderful ideas at wonderfulidea.co.

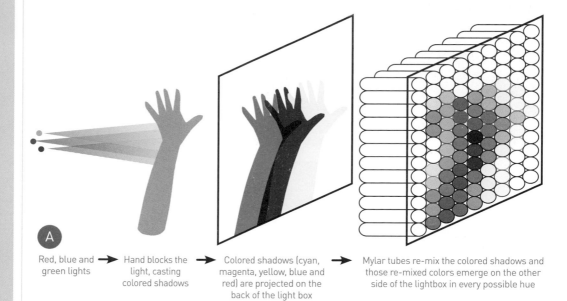

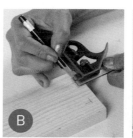

A

Red, blue and green lights → Hand blocks the light, casting colored shadows → Colored shadows (cyan, magenta, yellow, blue and red) are projected on the back of the light box → Mylar tubes re-mix the colored shadows and those re-mixed colors emerge on the other side of the lightbox in every possible hue

TIME REQUIRED:
1 Day
COST:
$100–$125

MATERIALS
- » **Pine boards, 16¼"×3"×¾"** (4)
- » **Wood screws, Phillips flat head #8, 1½" long, (8)**
- » **Acrylic, 17"×17"×⅛", (2)**
- » **Painter's tape**
- » **LEE Diffusion Filter, CL216, 21"×24" sheet cut down to 17"×17" (2)** I got mine from filmtools.com
- » **Scotch tape**
- » **Wood screws, Phillips head #8, ¾" long (27)**
- » **Mylar, 2mm** a 4'×25' roll should be plenty
- » **Wood dowels, 1" and 1.5" diameter, 10" long (2)**
- » **PVC pipe, 2" diameter, 10" long**
- » **Aluminum angle, 1½", 1½" long**
- » **RGB LED, 3-Up Cree high-power LED with jumpers, LEDs in series/Red XP-E2 LED, Green XP-E2, Blue XP-E2 #CUSTOM-3UP-CREE** ledsupply.com
- » **Machine screws, socket head, 4-40 thread, ¼" long (2)**
- » **Wood block, 4"×4"×¾"**
- » **LED driver, BuckPuck DC LED Driver, 700mA, wired, non-dimming** ledsupply.com
- » **Screw terminal**
- » **Female pigtail connector, 5.5×2.1mm jack**
- » **12V Transformer, Phihong Wall Mount, #12V-WM-1A** ledsupply.com
- » **P-strap, ¼"**

TOOLS
- » **Combination square, 6"**
- » **Pencil**
- » **Drill bits, #43, and ³⁄₃₂" and ⅛"**
- » **Hand drill**
- » **Countersink**
- » **Phillips screwdriver, #2**
- » **Plastic drill bit, ⅛"**
- » **X-Acto knife**
- » **Ruler**
- » **4-40 tap and tap handle**
- » **Cutting fluid**
- » **Hex key, ³⁄₃₂**
- » **Soldering iron**
- » **Small slotted screwdriver, ⅛"**

The Rainbow Lightbox uses an array of mylar tubes to create beautiful pixelated shadows from colored light. I was inspired by Taizo Matsumura's "Hikari no Hako" kit, which is a small paper box filled with mylar tubes. I created my own larger version with hundreds of mylar tubes and instead of relying on colored filters, as Matsumura's kit did, I used multicolored light and shadows to create a unique effect that brings out every color of the rainbow.

By experimenting with different light sources, from sunlight to televisions, I discovered that the mylar tubes do something surprising with colors if you sandwich them between two diffusion filters. When one end of a mylar tube is exposed to multiple colors, the tube mixes the colors together to create a new color that is projected onto the top diffusion filter. Although the colors that go into the tube array can vary smoothly, because of the sharp borders created by the edges of the tubes, the image that emerges has a beautiful pixelated appearance.

I also set about creating a colorful input to illuminate the lightbox, using red, blue, and green LEDs. When these colors are combined they create white light, but if you block them they cast colorful shadows onto the back of the lightbox. The colored shadows are re-mixed by the mylar array and emerge on the other side in every possible hue (Figure A).

The result is the Rainbow Lightbox: a device that lets you mix colored light like paint and play with color by casting shadows.

1. BUILD THE FRAME
On one end of each board, mark two screw holes that are ¾" in from each of the long edges, and ⅜" in from the short edge (Figure B). Use a ³⁄₃₂"

B

C

D

E

bit to predrill the screw holes. Countersink each hole (Figure C) — this will keep the wood from splitting when you screw the boards together. Line up the holes on one board with the undrilled end of a second board. Use a ³⁄₃₂" bit to carefully drill through the first hole into the end of the second board. Screw the boards together with a 1½" wood screw (Figure D). Repeat with the second hole. Repeat the whole process until you have a four-sided frame (Figure E).

Hep Svadja

Nicole Carrett

G

H

I

J

K

L

M

N

Nicole Catrett

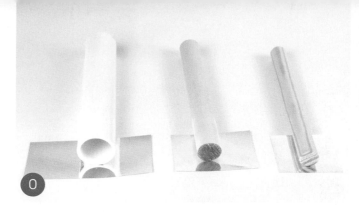

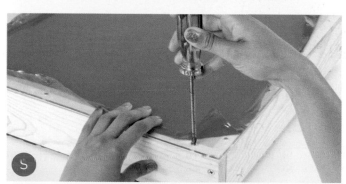

2. ADD THE ACRYLIC PANELS

Mark ⅜" in from the edge of each acrylic sheet, all the way around. Evenly distribute 16 screw holes around the edge of each panel, starting 1" in from each corner (Figure F). Use a ⅛" plastic bit to drill holes where the screws will go. Set your drill to a high speed and use low pressure to keep the plastic from cracking when you drill through. It also helps if there is something to drill into underneath the acrylic — a piece of plywood should do the trick (Figure G).

Place one of the acrylic sheets on top on the wood frame and secure it with painter's tape. Use the 3/32" bit to drill through each hole (Figure H). When you've drilled all the holes in one side, flip the frame over and repeat the process with the second acrylic sheet. Mark where each sheet lines up (Figure I — trust me, you'll never remember!) Remove the acrylic sheets, and set one aside for now.

Remove the protective covering from one side of the other acrylic panel (Figure J). Line the panel up with one of the diffusing filters, and use a bit of scotch tape to hold them together (Figure K). Carefully cut out the holes in the diffusion filter with an X-Acto knife (Figure L). Place the filter/acrylic sandwich back on top of the wood frame, filter side down. Peel the protective covering away from the screw holes and use ¾" wood screws to attach the acrylic panel to the frame (Figure M).

3. MAKE THE TUBES, LOTS AND LOTS OF TUBES!

Invite some friends over for this part with offers of cookies and beer; the work goes a lot quicker with good company! Cut the mylar into long, 3"-wide strips using an X-Acto knife and a ruler (Figure N). Lots and lots of strips. When you have a big pile of 3" strips, cut them down into 3½", 4½", and 6½" lengths (Figure O — the three lengths are for the different dowel diameters).

Now comes the cookies and beer part! Wrap a piece of mylar around a dowel to make a tube. Secure the mylar tube lengthwise with a strip of scotch tape (Figure P). Slide the mylar tube off the dowel and place it in the frame (Figure Q). Keep adding tubes until the frame is packed tight (Figure R). When you're satisfied, place the second diffusion filter on the filled frame, followed by the acrylic panel (aren't you glad you marked how it goes together?). Cut out the holes in the filter with an X-Acto, and screw it together (Figure S). Remove the protective coating from both sides of the lightbox (Figure T).

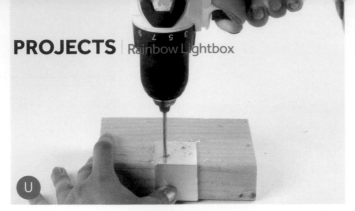

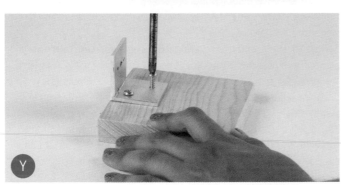

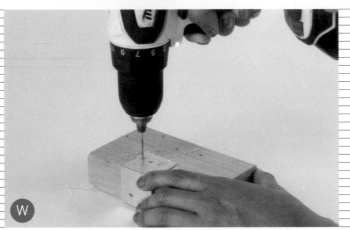

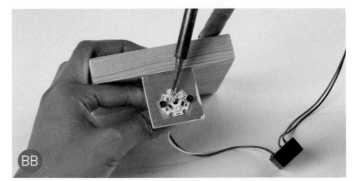

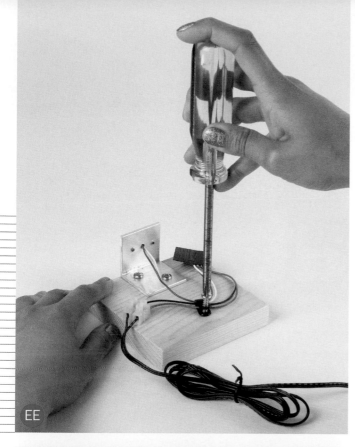

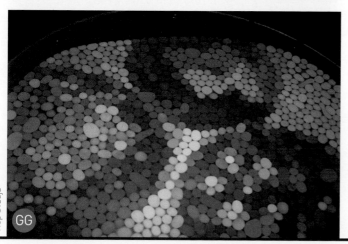

4. MAKE A HEATSINK

Super bright LEDs can put out a lot of heat (along with a lot of light), and it's important to give your LED a heatsink to dissipate that heat and keep your LED from burning out.

Use the ⅛" bit to drill a couple of screw holes on one side of the aluminum angle. It helps to put the angle on a block of wood for this part (Figure U). Place the LED on the other side of the angle, and mark the hole in the LED's center, and where you want the two 4-40 screws the go (Figure V). Drill the center hole with the ⅛" bit. Drill the two holes with the #43 bit (Figure W). Use the 4-40 tap and a little cutting fluid to thread the #43 holes (Figure X). Line the aluminum angle up with the edge of the 4"×4" wood block and screw it down using two ¾" wood screws (Figure Y). Put the LED in place and screw in the two 4-40 screws. Tighten them down with the ³⁄₃₂ hex key (Figure Z).

5. BUILD YOUR LIGHT SOURCE

Thread the BuckPuck's LED (+) and LED (-) wires through the center hole (Figure AA). Solder them to the LED's corresponding positive and negative pads (Figure BB). Use the slotted screwdriver to screw the BuckPuck's positive and negative wires into one side of the terminal block (Figure CC).

Plug the pigtail into the power supply jack and screw the positive and negative wires from the pigtail into the other side of the terminal block (Figure DD). Add a p-strap to hold everything in place (Figure EE). Plug in your light (Figure FF)! (If it doesn't turn on, double-check the polarity of the wires.)

USE IT

Find a dark space to play with your Rainbow Lightbox. Plug in your light, and shine it at the back of the lightbox (move the light around until you find the sweet spot where the whole lightbox is illuminated). Try casting colorful shadows by blocking the light with your hand. What happens if you move your hand closer or farther from the light? What colors do you see on the back of the box? What colors can you see on the front?

Try using different objects to create colored shadows, like lenses, water, mesh, crocheted lace, perforated metal ... look around you for things that might make interesting shadows or patterns (Figure GG). You could also try using different colors for the lights, or other light sources, like a sunny window or a television. If you want to go further, you could activate your lightbox with a mechanical shadow sculpture, turn it into a beautiful lamp, or make several lightboxes and stack them together to create an immersive environment. You can also play with the dimensions of your lightbox. I turned mine into a small table that several people can sit around to play with collaboratively. I think that what makes the Rainbow Lightbox so special is the way it reveals the beauty and physics of light through play and exploration. ◉

Hep Svadja

Text a Treat

This smart system dispenses snacks and sends you a pic of your pooch, all via SMS

Written and photographed by Rich Nelson

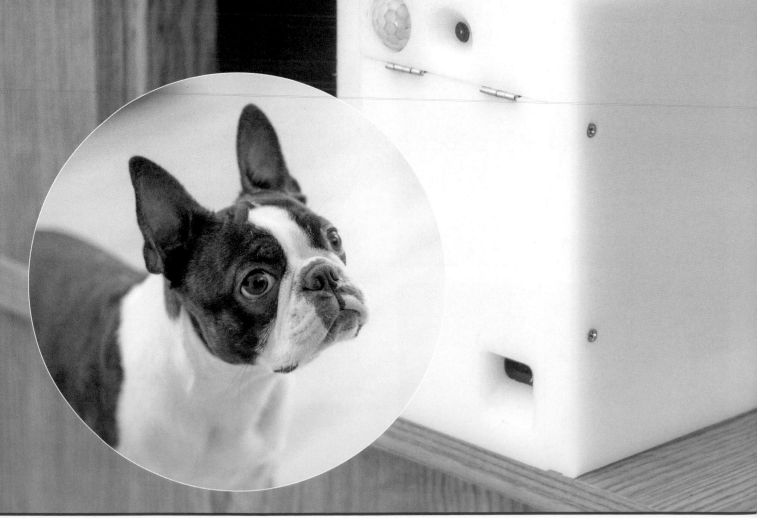

TIME REQUIRED:
1 Week

COST:
$40–$100

MATERIALS

» **Raspberry Pi, any model with a camera connector will work** such as v2, v3, Zero v1.3, or Zero W
» **Raspberry Pi Camera Module, V1 or V2**
» **Servo motor, SG90**
» **Power source, 5V, at least 2A**
» **Acetal plastic or alternate material** for enclosure and dispenser mechanism
» **Popsicle stick,** or other material for servo linkage
» **Flat head screws, #4×½" (16)**
» **Self-tapping screws, #2×5⁄16" (18)**
» **Disc magnets, ½"×1⁄16" (2)**
» **Butt hinge, ¾"×5⁄8" (4),** with 4 screws each
» **Jumper wires (9)**
» **Double-sided tape**
» **Dog treats** I buy Old Mother Hubbard Classic natural dog treats

TOOLS

» **CNC machine,** or 3D printer or woodworking tools
» **Computer with internet access**
» **Router with admin access (or a port tunneling service)**

RICH NELSON is an engineer by day at ROAR for Good, using technology to make the world a safer place. By night he works on projects of limited use and maximum amusement. Visit RichNelson.me to see more of his projects.

Leaving our new pup Gus at home was "ruff," especially for my wife. As a maker I saw a problem that needed solving, and the approaching holiday season set a deadline for the perfect gift. The project quickly took form: find a way to check in on the dog while we were at work and maybe even interact with him a bit.

I knew I could use a Raspberry Pi Camera Module to take photos and control electronics. For connectivity I'd seen a few projects using Twilio, a service for internet-connected, programmable phone numbers. This was a perfect fit — Gus could literally send us a text! Lastly, I needed a way to get Gus's attention for the photos ... treats were the obvious answer.

For my second Raspberry Pi project, this seemed like a big challenge. But like any complex task, I broke it down into manageable pieces and solved them one by one. This made it easy to get started, and each small win pushed me forward. Like John Wanamaker said, "One may walk over the highest mountain one step at a time."

While I made a few upgrades along the way, the original setup goes like this: I send a text to my dog's Twilio phone number. Twilio forwards the text to the Raspberry Pi server. The Raspberry Pi uses the Arduino to wiggle the servomotor in order to get Gus' attention. The Raspberry Pi takes a photo, and a treat pops out for Gus while the photo is uploaded to Dropbox. Finally, Twilio texts me the photo. To a dog, it's a magical white box that gives treats. To a human, it's a way to check in on your pup and get a cute pic when you're out for the day.

SET UP TWILIO

Twilio is a game changer. Inexpensive electronics allow anyone to make awesome devices at home, and now with Twilio we can connect with those projects through text messages or phone calls.

Uber uses Twilio to text you when your ride is coming. The startup that I work for, ROAR for Good, uses it to notify your loved ones when you are in danger. With a few lines of code and less than one cent per text, we makers can extend our projects beyond our households with no custom mobile app or web interface required.

Twilio provides a unique phone number. When a text is received, it triggers an HTTP request. This is just like typing a

Treat

web address into a browser, but instead of requesting a webpage from the server it sends the info from the text message (Figure). Twilio supports Python, which is native to all Raspberry Pi boards. To get started, check out Twilio's guide at twilio.com/docs/guides/how-to-receive-and-reply.

With just 11 lines of code I had a simple web server that could receive and automatically respond to text messages! Before going live I had to expose the server to the internet with port forwarding on my router. With the basic server done, the project was starting to take shape, and I continued to add functions as the remaining features were completed. You can find the finished code on my GitHub at github.com/rmn388/dog-treat-dispenser.

DISPENSING THE TREAT

For the mechanical dispenser, I took

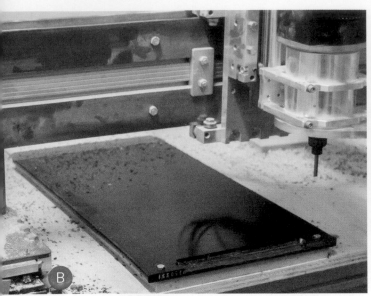

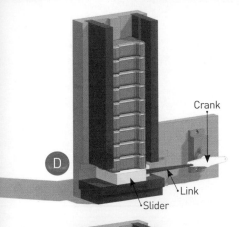

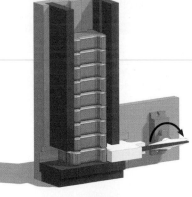

inspiration from a gun magazine (with tasty treats in place of bullets). I measured some of our dog's favorite treats, and designed the channel around them. I CNC cut the dispenser assembly (Figure B and top right in Figure C) out of acetal plastic.

I designed the dispenser around a common servo motor to keep it affordable. I converted the servo's rotational motion into linear motion using a crank, link, and slider mechanism. This is similar to a crank and piston in a car engine. In this case the white plastic servo arm acts as the crank. I cut up a popsicle stick for the link, and the slider is a machined plastic piece (Figure D).

Servos are driven by PWM (pulse-width modulation) signals, essentially electrical pulses that vary in length to control the angle of the servo arm. To keep the hardware as simple as possible I attempted to use a Raspberry Pi to drive the servo. The Pi does not have native PWM hardware, but there are some software libraries that can emulate this signal. Unfortunately these were buggy in my experience, leading to jittery, unreliable servo movement. As a quick solution I wired up an Arduino for servo control and it worked perfectly. This highlights one of the primary differences of an Arduino (microprocessor) vs. a Raspberry Pi (single board computer). Arduino can only do one thing at a time, which is great for extremely time sensitive actions like servo control. A Raspberry Pi has a full operating system running, so other processes can interrupt your program at any time.

WORTH A THOUSAND WORDS (OR ONE TREAT)

After plugging in the camera and installing a few packages, it only took two lines of code for the setup, then one more line to take photos! This was by far the easiest part of the project because the Raspberry Pi system is so well supported.

```
import picamera
camera = picamera.PiCamera()
camera.capture('image.jpg')
```

That saves the file to the local folder, but Twilio needs a URL in order to text the photo. Luckily Dropbox provides a simple Python API to upload a photo and get a shareable link. As a bonus, this also saves the photos in a shared Dropbox so my wife and I can browse the photos from anywhere (Figure E).

THE ENCLOSURE

I chose acetal plastic for the box as well (shown in white in Figure C), hoping the tough plastic would be chew-resistant against our persistent pup. The CAD files are also on my GitHub. I machined the parts on a CNC router, but a comparable enclosure could be made with 3D printing, traditional woodworking methods, or whatever tools you have at your disposal.

All in all the design is meant to be unassuming so it doesn't stand out in our

Crank

Link

Slider

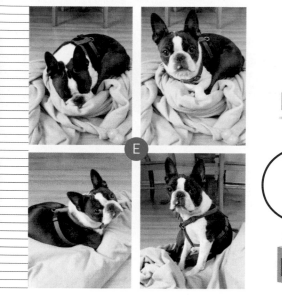

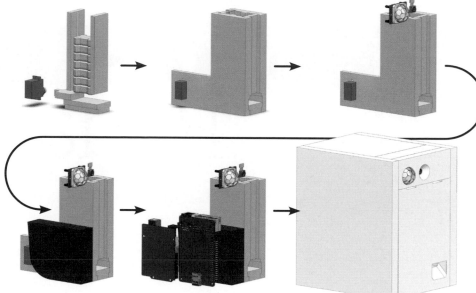

house. Though, one notable feature is the camera tilt mechanism. Using a live hinge (a thin bendable section of the plastic), the camera can be tilted up and down by sliding the lid forward and backward.

UPGRADES

This dispenser has been useful in our day to day, but like any project it's not always used exactly as you imagine. With all the treats, Gus started packing on the pounds! To keep him trim, I changed the default behavior to stealthily capture a photo. Now we only send the occasional treat with a specific text command. I extended the functionality to automatically send a photo every day at noon, which is nice to get in the middle of the day. I also added a feature so my wife and I could have the captured photos sent to each other.

I have a few ideas to take this project to the next level. Adding live streaming video would be cool. Or adding a motorized pan/tilt mechanism, or turning it into a remote control vehicle that could drive around to find Gus. I wired up a motion sensor as well. Although it's not currently being used, it's ready to be enabled through software if we want to automatically take photos when Gus moves around, or use the OpenCV computer vision library to detect if he is in frame. ◗

Check out makezine.com/go/treat-dispenser to see the treat dispenser in action and to get more info.

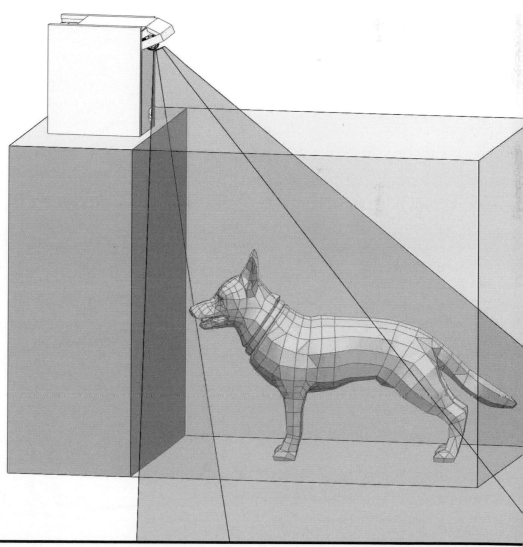

Robot-Ready
Radar

Written by Adam Kemp

An ultrasonic sensor and a single servo let your bot sense its surroundings for less than $10

TIME REQUIRED:
30–60 Minutes +
3D print time

COST:
$5–$10

MATERIALS

- » **Sensor holder** download the files from Thingiverse thingiverse.com/thing:2481918
- » **Ultrasonic sensor, HC-SR04 or equivalent** RobotShop #RB-Lte-54 robotshop.com
- » **Micro servo motor** #RB-Dfr-124 robotshop.com
- » **Servo extension cable, 30cm** #RB-Dfr-179 robotshop.com
- » **Screw, M3 or smaller** (4)
- » **Cable tie** aka zip tie
- » **Arduino Uno or similar**
- » **Robot platform, two-wheel, continuous rotation servo-driven**

TOOLS

- » **3D printer** You can print the parts yourself, or send the 3D files to a service for printing. See makezine.com/where-to-get-digital-fabrication-tool-access to find a printer or service you can use.
- » **Repo for ultrasonic library** at github.com/ErickSimoes/Ultrasonic
- » **Soldering iron** adjustable temperature with stand
- » **Solder** standard rosin core
- » **Heat-shrink tubing, 1/16"–1/8"×2"**
- » **Snips** small or flush cut
- » **Strippers, 20–30 gauge**
- » **Screw driver, #1 Phillips**
- » **Hobby knife**

Make: author **ADAM KEMP** has taught high school technology and engineering for over 12 years and currently co-chairs the Science, Technology, Engineering, Art, and Mathematics department at the Princeton International School of Mathematics and Science. His book, *The Makerspace Workbench*, has served as a resource for thousands.

A utonomous vehicles use a wealth of sensing technologies to calculate relative position and make driving decisions. There are sensors that determine geolocation, cameras for real-time image processing, and others that measure speed and magnetic heading, yet the one that I find the most interesting is the lidar. Similar to radar, lidar continuously sweeps a high speed, laser-based distance sensor to generate either a 2D or 3D point map of the surrounding landscape. This massive amount of information is then processed and adjustments are made to the trajectory of the vehicle. Unfortunately, rotating lidar systems are very costly and require significant processing capability to fully utilize. Luckily the same principles can be applied on a smaller scale using a conventional ultrasonic sensor and a single servo.

Are you ready to give your two-wheeled Arduino-controlled robot some cool, smart-sensing abilities? Great! Gather all of the tools and materials (Figure 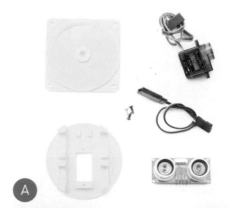) you need and find out how for less than $10 you can create a sweeping 3D printed ultrasonic sensor aptly named ultradar!

1. PREPARE THE SENSOR

Start this project by visiting my Thingiverse page at thingiverse.com/thing:2481918 and downloading the latest files for the ultradar assembly. The download includes the solid models and software to run your robot. Print the top and bottom models using the following settings:

Plastic Type:	PLA or ABS, others plastics are untested
Layer Height:	0.1mm - 0.2mm
Infill:	No less than 10%
Support:	Touching build plate
Brim:	No, unless you have adhesion issues

While the assembly is printing, modify the sensor by soldering the Trig and Echo pins together (Figure ). This will allow you to use the sensor in three-wire mode and can therefore control it with a repurposed servo cable. Trim off any excess material and shorten the pins by 2–3mm (Figure ).

Remove the male end of the servo extension using your snips. Strip about 3mm from the end of each wire and tin with solder. Separate roughly 2cm–3cm of each wire from the bundle and slide a 1cm piece of heat-shrink tubing over each (Figure ).

Solder the red wire to 5V pin, the white/orange wire to the Echo/Trig pin, and the black/brown wire to the Gnd pin (Figure 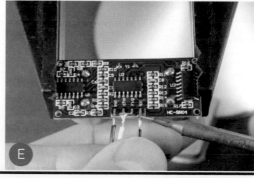). Be careful not to

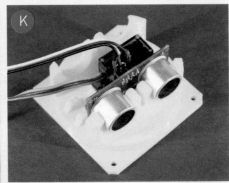

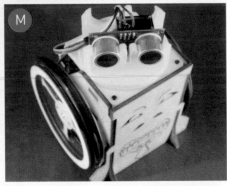

```
                 -5v  -50.00ms
0-25, 45-76, 90-37, 135-14, 180-236, max = 180
Left: 40.00ms
0-99, 45-224, 90-31, 135-33, 180-266, max = 180
Left: 40.00ms
0-78, 45-139, 90-5, 135-7, 180-200, max = 180
Left: 40.00ms
0-122, 45-194, 90-29, 135-77, 180-78, max = 45
Right: 20.00ms
0-99, 45-112, 90-211, 135-34, 180-102, max = 90
Forward
0-75, 45-167, 90-79, 135-10, 180-100, max = 45
Right: 20.00ms
0-105, 45-119, 90-16, 135-19, 180-101, max = 45
Right: 20.00ms
0-118, 45-183, 90-46, 135-19, 180-103, max = 45
Right: 20.00ms
```

overheat your wire as it will cause the heat-shrink to shrink.

Slide each piece of heat-shrink over the soldered joints. Make sure that all of the exposed wire and pin are covered. Use a heat gun or lighter to shrink the heat-shrink around the solder joints (Figure F).

2. ASSEMBLE THE ULTRADAR

Remove the support material from the screw hole located on the base piece and test fit the servo screw (Figure G). Set aside.

Clean up any remaining support material from the top piece and make sure that the holes for the sensors are free of burrs. Carefully press the HC-SR04 sensor into one of the two mounts until the crystal just touches the plastic support (Figure H). Do your best not to dent the sensor's aluminum cylinders.

Insert the servo into the mount so that the output shaft is centered (Figure I). Secure in place with the two mounting screws provided with the servo (Figure J).

Upload the "ServoTest" program provided with the Thingiverse ultradar files to your Arduino and attach the servo cable to D10. The program will start by outputting a servo command of 90 and will center the servo. Now that the servo is centered, attach the bottom half of the ultradar assembly using the appropriate screw. Be sure to align the top and bottom halves so that the front of the sensor is parallel to the bottom (Figure K). Once secured, you can disconnect the servo from D10.

Route the sensor and servo wires through the cable mount and tighten in place using a cable tie (Figure L). Trim off the excess material using your snips. This mount will help prevent the servo and sensor wires from fatiguing over time.

Mount the ultradar sensor onto your robot so that the center position is directly forward (Figure M). Connect the servo to D11 and the sensor cable to D13 and you are ready to test!

Upload the "UltraDAR-SingleSweep" program to your Arduino and open the Serial Monitor. The sensor should quickly sweep 180° and the readings will be printed to the monitor (Figure N). If you do not see the data stream check that you have connected the sensor to D13 and the orientation is correct.

Adam Kemp

3. PREPARE YOUR ROBOT

You will need to trim your continuous rotation servos before you run the robot. This will ensure that the robot drives as straight as possible and the adjustment turns are accurate. Reload the "ServoTest" program and attach one of the continuous rotation drive servos to D10. If the servo begins to rotate, use a small Phillips screwdriver to adjust the trim potentiometer until the servo stops (Figure O). This will set the servo's physical midpoint.

Record the software midpoint by manually adjusting it through the Serial Monitor. Open the Serial Monitor, set the baud rate to 9600, and send a "-" to decrease the servo control number until the service begins to turn (Figure P). Record the number. Now increase the number by sending an "=" until the servo begins to spin in the opposite direction and record the number. Average the two numbers to determine the servo's software midpoint. Repeat the previous two steps for the other drive servo.

Disconnect the servo and connect the left drive servo to D10 and the right to D9. Double-check that the sensor is connected to D13 and its sweep servo connected to D11. Open the "ultradar-SingleSweep" program and update the "leftCenter" and "rightCenter" constants to reflect your calculated software midpoints (Figure Q).

The remaining constants determine how the robot will respond to the surrounding environment. You can leave them as-is or adjust them based on the following table. The min and max values should be acceptable for a non-geared two-wheeled servo driven robot.

4. UPLOAD AND TEST

Save and upload the modified code to the robot. The servos should twitch and the sensor sweep servo will move to its home position. After 1 second, the sensor will begin to sweep and the wheels will turn accordingly.

Place the robot in an open area and off it will go! If everything is installed correctly and set properly, the robot should autonomously navigate the surrounding area while endlessly trying to find the longest path to follow.

A SENSATIONAL STEP

Congratulations! With the addition of an ultradar sensor, your robot has taken one step toward full autonomy. Spooky, right? Try running it through a maze made out of books, or put it outside and see how far it can drive before it gets stuck. The possibilities are endless! I hope you have enjoyed this project and explore the many ways it can be used to give your robots a better sense of their surroundings. ⊘

CONSTANT	FUNCTION	Min Value	Max Value
leftTurnTime	Delay in ms per left turn step	10	100
rightTurnTime	Delay in ms per right turn step	10	100
reverseTime	Time in ms for reverse action	100	1000
reverseThreshold	Distance in cm to trigger reverse	5	15
maxSweepAngle	Max angle in degrees for sensor to sweep	0	180
sampleAngle	Angle at which each measurement is taken. Should be a divisor of maxSweepAngle.	15	90
sampleDelay	Delay in ms after each distance measurement.	0	10
numSamples	Quantity of measurements averaged at each sample angle	10	30

Written by Brookelynn Morris

Kokedama String Garden

Transform your plants into hanging sculptures

TIME REQUIRED:
20–30 Minutes
COST:
$10–$20

MATERIALS
» **Small plant**
» **Twine, heavy-duty**
» **Heavy clay mud, or bonsai soil**
» **Potting soil**
» **Sheet moss**
» **Sphagnum moss (optional)**
» **Peat soil (optional)**

TOOLS
» **Scissors**

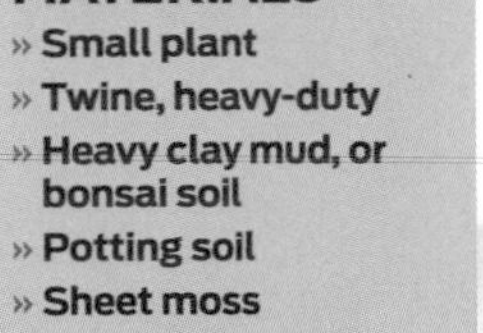

BROOKELYNN MORRIS is an OG maker living in the redwood forest on the coast of Northern California. Find her on Twitter @Brookelynn23.

Nat Heckathorn

Being a bonsai and ikebana enthusiast, I was captivated the first time I saw a *kokedama* (moss ball) string garden. I've hung my kokedama outdoors, where it sways a bit in the breeze, and I couldn't love it more.

The steps are simple, with the most important aspect being the mixture of clay and soil. Try to achieve a blend of mud that clings together well. When encasing the plant, you will feel a bit like a sculptor, and in essence, you are sculpting the earth into something fairly impossible — a perfectly round hanging garden.

1. COMBINE NUTRIENTS
Mix a few generous handfuls of clay with an equal amount of potting soil (Figure Ⓐ).

2. REMOVE PLANT FROM CONTAINER
Pull the plant from its container (Figure Ⓑ). To trim the roots, tug at them gently, breaking them apart, without breaking them off (Figure Ⓒ). Rinse as much of the growing media from the roots as you can.

3. CREATE SOIL BALL
Hold the plant with one hand, and with the other, pat the clay mixture all around the roots. I started at the sides, worked to the top, then pressed the clay into the bottom of the plant last. Mold the clay into a round ball, making sure that the mud is against the base of the leaves (Figure Ⓓ).

4. ATTACH MOSS
Firmly press moss all over the clay ball. The soft moss should stick right to the mud (Figure Ⓔ).

5. WRAP IN TWINE
Cut at least 3 yards of twine. Lay the moss-covered plant onto the center of the twine, then wrap and twist (Figure Ⓕ), wrap and twist, wrap and twist (Figure Ⓖ) until the entire garden is bound.

6. CREATE THE HANGER
To finish the wrapping, draw one of the ends of twine up the side of the ball. Weave it over and under the wrapped twine, knotting as you go. Then repeat with the other end of string, on the opposite side of the ball. Getting a nice balance is crucial, especially with an upright growing flower (Figure Ⓗ).

7. DISPLAY
Hang the kokedama where the conditions are suited to the plant's needs (Figure Ⓘ). To water it, hold a container of water up to the wrapped ball to submerge the roots. Because the plant is not in a solid vessel, it will require a bit more frequent watering than other plants in your garden.

GOING FURTHER
Your kokedama options are boundless. Hardy plants that require little root space work best. For additional moisture, you can wrap the plant's roots in Sphagnum moss before creating the soil ball. Mixing peat, bonsai, and potting soil together will up the nutrients as well as enhance the kokedama's moisture retention. ✪

Fat Head

Supersize your noggin with a lens and a cardboard box

Written by Yuji Hayashi with Tomofumi Yoshida and Rei Betsuyaku

TIME REQUIRED:
2 Hours
COST:
$30–$40

MATERIALS

» **Cardboard box** Any box is fine as long as it's about 15" deep. We used a document holder 120 size (about 15"×12½"×11½")

» **Fresnel lens** from Aliexpress. Things looked interesting with a focal length of 500mm.

» **LED tape, warm white** The same type used for cars or interior lighting. We used a 500 LED strip. You can use other colors, but your face won't be as clear. Make sure you choose LEDs in the 3,000K–4,000K range so the light appears "natural."

» **Batteries and holder** paired to your LED tape's working voltage. Our LEDs came with one already attached.

» **Wire, two colors** to connect the LED tape and the battery box

» **Heat-shrink tubing**

TOOLS

» **Box cutter or scissors** to cut the cardboard

» **Adhesive tape** to affix the lens to the cardboard. Transparent tape looks better.

» **Soldering iron and solder** to attach the LED, electricity, and cord

» **Plastic cutter (optional)** to cut the Fresnel lens

» **Double-sided tape (optional)** if LED tape doesn't come with an adhesive backing

» **Sturdy gloves (optional)** to protect your hands if cutting the lens

» **Snacks (optional)** eat these when you need a break

YUJI HAYASHI
is the editor-in-chief of Daily Portal Z (portal.nifty.com).

Hep Svadja, Daily Portal Z - portal.nifty.com

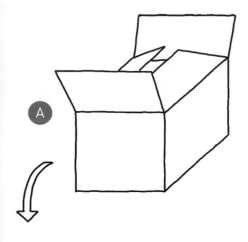

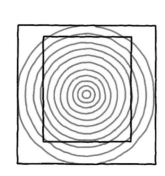

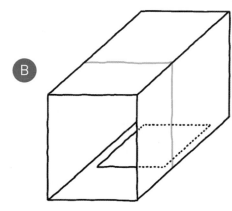

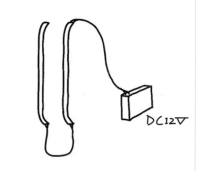

We held the Big Face Mask Workshop at Maker Faire Tokyo in August 2016. Originally, we were going to have participants make a large version of their face using papercraft. However, it took 2 hours per person. As a way to make it much simpler, we decided to ask them to wear a box instead. Thanks to this idea, it only takes 2 seconds now with the Bigfacebox. I have to wonder if I can still call this a workshop.

This is not the only way to do it. Please feel free to choose your box size, lens type, lighting, and such to get your desired result.

> ⚡ **CAUTION:** Please do not look directly into the sun or other bright lights when wearing the face enlargement box.

1. ASSEMBLE THE BOX

Close one set of flaps. Tape the other set of flaps open so that they extend the height of the box (Figure A). Then lay the box down so the opening is on the side.

Make a hole on the bottom where you can insert your head (Figure B). Positioning your head about 6" away from the opening makes for an interesting face.

2. FIT THE LENS

Depending on the size of your Fresnel lens, you may need to cut it to fit the opening of your box using an acrylic cutter. Please be careful not to cut your hands when cutting the lens' cross-section. (I cut mine three times. Work gloves are highly recommended.)

If you get a rectangular lens rather than a circular lens you will have less work to size it to your box opening. Make sure you preserve the center of the lens. If you need to cut off two inches, cut one inch off each side (Figure C).

Your face will look more interesting when the lens' center is set slightly back from the opening.

3. ATTACH THE LED STRIPS

Cut two lengths of the LED strip that measure slightly less than the long edge of the lens. Our lens was 15"×12½", so we made the length of the strips less than 14". Then measure two lengths of wire slightly longer than the short edge of the lens, and solder together the two ends of the LED strips (Figure D), one color to the positive pads and the other to negative pads. Cover the solder joints with heat-shrink tubing to protect against pulls and shorts. The wire to the battery holder should be long enough to be out of the way.

Attach one strip of LED tape on each long edge of the lens' ridged side (Your face will look better if the smooth side of the lens faces out). LED tape with adhesive on the back is most convenient, but you can use double-sided tape, too. (You can also put an LED strip on the top and bottom, as shown in the photo.)

4. PUT IT ALL TOGETHER

Attach the lens to the box with adhesive tape, then attach the battery holder to the inside of the box. Turn on the LEDs, put the box over your head, and it's complete (Figure E). Let's scare some people. ❷

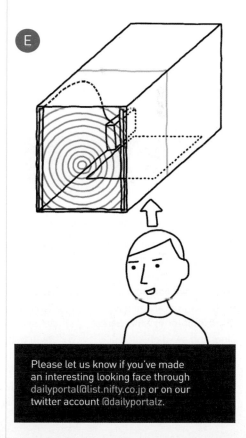

> Please let us know if you've made an interesting looking face through dailyportal@list.nifty.co.jp or on our twitter account @dailyportalz.

CAMILO PARRA PALACIO is an Industrial Designer in Shanghai, China, creator and founder of Otto DIY, a project that follows his passion for robotics, toys, open source hardware and his dream to be a Maker Pro.

DIY Bipedal Robot

Open source and customizable, Otto can dance, make sounds, and avoid obstacles

Written by Camilo Parra Palacio

Otto is an open source, Arduino-compatible, 3D-printable bipedal (two-footed) robot made with off-the-shelf components. It can be programmed from any computer to walk, dance, sing, and avoid obstacles. Standing a little under 4½" tall and with fewer than 30 parts, Otto is the perfect platform for working with a robot that you build yourself, and makes learning about programming and robotics easy, interactive, and fun!

EDUCATIONAL

Otto is also a great tool for the classroom, and it's currently being used by educators around the world to teach children (ages 8 and up) about the wonders of robotics. Adults are, of course, joining in the fun as well.

Otto follows a line of microcontroller-powered, do-it-yourself, bipedal robots. Thanks to the open source nature of these projects, each generation is able to build upon its predecessors (Figure Ⓐ).

The Otto project is licensed under the Creative Commons Attribution-ShareAlike 4.0 International (CC-BY-SA). This license lets others remix, tweak, and build upon your work even for commercial purposes, as long as they credit you and license their new creations under the identical terms. For the full legal code of this license please visit creativecommons. org/licenses/by-sa/4.0/ legalcode.

PROGRAMMABLE

Once assembled, Otto can be easily programmed over USB to perform several different

Otto DIY

Solar-powered driveway lights seemed like a great idea when I installed them two years ago, but their transparent plastic turned yellow in the sun, and their batteries wore out. I decided to scrap them and install 40 LEDs wired in parallel. If I used an adjustable supply to provide the recommended forward voltage, I wouldn't need series resistors (see Figure Ⓐ).

I will swallow my embarrassment, now, as I explain how wrong I was.

I have a long driveway, but I determined that 800 feet of 18-gauge wire would have a resistance of only 5 ohms. By comparison, if an LED passes 20mA at 3.2V, it has an effective resistance of 3.2 / 0.02 = 160 ohms, according to Ohm's law. Of course, an LED doesn't have resistance in the formal sense, because it varies with voltage. But the calculation gave me a ballpark figure to reassure me that the resistance of the wire should be trivial by comparison. (If you're sketchy about Ohm's law, you'll find it fully explained in my book *Make: Electronics*.)

I set to work pulling wire and burying conduit. At the end of the day it suddenly occurred to me (duh!) that if I had 40 LEDs in parallel, their collective resistance would be around 4 ohms, not 160. In which case — maybe the resistance of the wire would not be trivial, after all?

When in doubt, do the math. Figure Ⓑ shows four lights wired in parallel. WR is the wire resistance of each section, while AR1, AR2, and AR3 are adjustment resistors chosen to supply the same current to each light. What should the resistor values be? By applying Ohm's law and thinking very hard, I derived the formula shown. Figure Ⓒ shows how to verify it yourself, if you wish.

ASK AN EXPERT

My formula told me that using 10-foot sections of 18-gauge wire, AR1 through AR40 would range from 0.1 ohms all the way up to 105 ohms. Such a high value seemed odd, so I decided to consult my friend Ken, who has decades of experience as an electronics engineer.

"But you shouldn't be wiring the LEDs in parallel," he said.

Really? How else should I wire them?

"In series, with a constant-current power supply."

A what?

Kindly, Ken elucidated. The behavior of

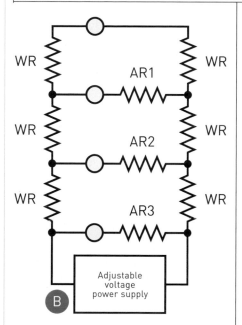

Ⓐ When lights are wired in parallel, each receives the same voltage (ignoring distribution losses caused by wire resistance) but some lights can draw more current than others.

Ⓑ Lights wired in parallel, with sections of wire having resistance WR. Adjustment resistors AR1, AR2, and AR3 compensate for wire resistance so that the same current passes through each light. It can be proved that

AR1 = 2 × WR
AR2 = (4 × WR) + (2 × WR)
AR3 = (6 × WR) + (4 × WR) + (2 × WR)

...

ARn = n × (n+1) × WR

The resistance of each light is unimportant, so long as they are all equal.

To apply a specific voltage across each light, adjust the power supply.

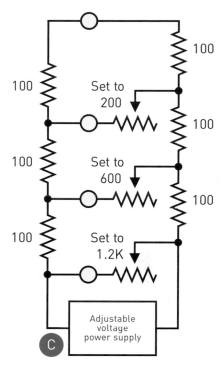

Ⓒ To verify the formula from Figure B, use 100Ω resistors to represent each segment of wire and 2K trimmers as adjustment resistors. Any resistance may be used to represent each light, so long as they are all the same. Voltage drop and current will be equal for each light.

an LED is affected not only by voltage but also by manufacturing inconsistencies. Consequently, a voltage that is correct for most will be a little too high for some. This greedy minority will suck more milliamps than they should, curtailing their own life expectancy.

I found this hard to believe, so I tested some LEDs at their recommended 3.2V forward voltage. Sure enough, I found one that gorged itself with 31mA, exceeding the absolute maximum specified in its datasheet.

This kind of problem doesn't arise when you're building a small device that uses an LED as an indicator. Just add a series resistor of the next-higher standard value, and if the LED is slightly underpowered, it will still be bright enough to serve its purpose.

When LEDs are used for lighting, the situation is different. We want maximum brightness, and the best way to achieve this is by ordering a constant-current power supply from eBay for around $25. Set the supply for 20mA, apply it to LEDs

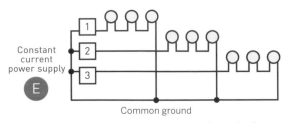

D Constant current power supply

When lights are wired in series with a constant-current power supply, an equal, controlled current passes through all of them.

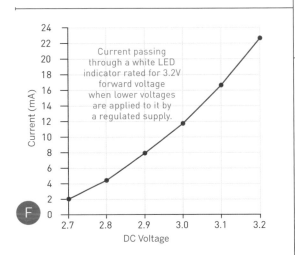

E Constant current power supply

1
2
3

Common ground

Multiple serial loops may be necessary if the total voltage required by lights in series exceeds the specification of a single constant-current power supply.

F Current passing through a white LED indicator rated for 3.2V forward voltage when lower voltages are applied to it by a regulated supply.

Current (mA) vs DC Voltage

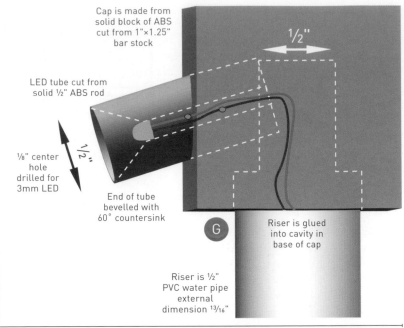

Cap is made from solid block of ABS cut from 1"×1.25" bar stock

½"

LED tube cut from solid ½" ABS rod

⅛" center hole drilled for 3mm LED

½"

End of tube bevelled with 60° countersink

G

Riser is glued into cavity in base of cap

Riser is ½" PVC water pipe external dimension ¹³⁄₁₆"

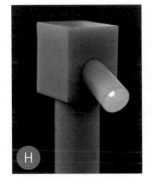

H

I

wired in series (as in Figure **D**), and every LED will deliver its maximum output without hurting itself.

Unfortunately for me, this created a new problem. If I put 40 LEDs in series, I would need 40 × 3.2 = 128V to power them all. Most constant-current supplies can't handle that, so I would need three supplies, each powering a separate loop, as in Figure **E** (which shows small groups for simplicity).

ASK ANOTHER EXPERT

I imagined digging up the conduits, ungluing them, desoldering the wires, adding new wires, and burying everything again.

I did not want to do that.

Well, if I didn't like Ken's expert advice, maybe I should try a different expert. I contacted my friend Graham, who is another seasoned electronics engineer.

"Oh, keep them wired in parallel," he said. "But use series resistors."

He reminded me that a series resistor doesn't just impose a voltage reduction. It functions as a current limiter, because of the way that an LED passes much more current when the forward voltage increases slightly. An increase of just 0.5V can allow current to increase by more than a factor of 10. This is illustrated in Figure **F**.

If you pair a resistor with the LED, any increase in current causes the resistor to impose a larger voltage drop, because voltage = current × resistance (Ohm's law again). The voltage drop causes the LED to pass less current, and somewhere in the middle, the two components find a stable equilibrium.

"Install 470-ohm resistors," Graham advised me. "Then increase the voltage of your power supply till the nearest LED is drawing 20mA."

Shouldn't I use trimmers so that I could adjust them to compensate for the wire resistance?

"I wouldn't bother. The LEDs farther away

will be a bit dimmer, but if you can't see the difference, don't worry about it."

WEIRD AND WONDERFUL

I did as he said, and sure enough, the LEDs do look okay. Of course, they are still wired in parallel, and even with their protective resistors, maybe one or two will overdose on amperage and die young. Only time will tell.

Bearing in mind all this aggravation, perhaps I would have been smarter to use store-bought lighting. But store-bought products are boring. Figure **G** shows the enclosure that I designed for each LED, Figure **H** shows a finished sample, and Figure **I** shows the front path to my house.

You must agree, you'll never find weird-looking creepy-green miniature lights like these at The Home Depot. ⊘

A Clean Sweep

Written and photographed by Sean Michael Ragan

TIME REQUIRED:
1–3 Hours

COST:
$0–$50

Light up your broom with LEDs from a scavenged worklight

MATERIALS
- » **Rechargeable strip LED worklight** I used 2, but 1 is sufficient
- » **Insulated stranded hookup wire, 22 ga** (48" each of red and black)
- » **LED flashlight, 3 x AAA** I used a 130 Lumen LED Flashlight from Defiant
- » **Blind rivets, ⁵⁄₃₂" diameter × ½" length** (3)
- » **AAA batteries** (3)
- » **Broom** I used a two-in-one angle broom/dustpan set from Vileda because it has lots of features that make this hack easier.
- » **Electrical tape or heat-shrink tubing**

TOOLS
- » Drill and bits
- » Round "rat tail" file
- » Razor saw or hacksaw
- » Pop rivet tool
- » Soldering stuff
- » String or twine (6')
- » Wire stripping/cutting pliers
- » Screwdriver set

SEAN MICHAEL RAGAN is descended from 5,000 generations of tool-using primates. Also, he went to college and stuff. Sean is the author of *The Total Inventor's Manual* (Weldon Owen, 2016).

My office looked clean enough under ordinary lighting, but with photo lights for a project shoot I discovered some neglected corners to be harboring colonies of dust bunnies. I found myself many times peering under a desk or behind a cabinet with a flashlight in one hand and a broom in the other. Clearly there was a better way …

For complete instructions, visit this project online at makezine.com/go/build-an-led-broom, but here are the basics:

❶ TAKE EVERYTHING APART
Keep the 6-diode white LED strip and 4 case screws from the worklight, and everything from the flashlight except the reflector, lens, and LED, which you will need to gently pry out and desolder or cut the exposed leads connecting it to the PCB underneath. Carefully drill a ¼" hole in the center of the lower end of the handle and a matching hole that lines up at the bottom of the handle socket.

❷ MOUNT THE FLASHLIGHT
Carefully file the retaining lip from the front end of the optics ring. Screw the flashlight body into the ring, and push it over the broom handle until it bottoms out. Mark three ⁵⁄₃₂" holes evenly spaced around the circumference of the ring and remove the flashlight. Drill all the way through the handle, then install a rivet in each hole, saving one of the rivet mandrels.

❸ RUN THE WIRING
Cut a 48" length each of red and black wire and tie them together in a knot about 4" from each end. Strip one end off each of the black and red leads and solder to the flashlight PCB pads, black to negative, red to positive. Tape or heat-shrink the other ends together and attach a 6' piece of twine. Tie the other end of the twine to the leftover mandrel and snake the twine down the length of the handle to the hole. Pull the twine and the wires through, take up the slack, then thread the flashlight body into the optics ring to secure it at the end of the handle. Stuff the free end of the wiring harness through the hole in the whisk socket, then screw the whisk to the handle. Cut off the twine and the tape or heat-shrink.

❹ INSTALL THE LEDS
Drill four ¹⁄₁₆" holes through the LED strip PCB between the five central diodes. Align the strip at the top of the bristles so that it's angled toward the floor. Use the PCB as a template to drill four shallow ¹⁄₁₆" mounting holes and secure with the 4 small case screws. Run the red and black leads through the factory mounting hole at the end of the PCB nearest the contact pads, then trim and strip the leads before soldering them to the pads: red positive, black negative.

USE IT
Load batteries into the flashlight body and click the button to illuminate the dark corners of your world. ⊘

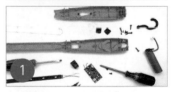

How to Build a Crappy Robot

STEP 1: Embrace Failure

The point of building a terrible robot is to have fun while learning and to enjoy the bumps and bruises along the way. The best way to do this is to make your task impossible. This takes off the pressure to succeed.

STEP 3: Make It Think!

STEP 4: Tell It When to Do Stuff!

Inputs:
» Voltage
» Keyboard input
» Sensors:
light, sound, movement, moisture, etc.

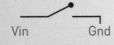
BRAINS!

Analog
Voltage ranges between 0V and whatever voltage your system is running at
(behaves like a dimmer switch)

Vin   Gnd
       Sig

Digital
Only have 2 states:
On or Off
(behaves like a regular switch)

Vin   Gnd

Once a signal passes through a sensor, it goes back into your microcontroller which then decides what to do with that information. In our case, this can mean triggering an output.

Microcontroller
To execute your commands:
» Arduino
» micro:bit
» Makey Makey
» Raspberry Pi
Use this criteria to find the best one for you:

Onboard sensors?
Easy to use and don't take up any extra space.

Language?
Can you choose what language is most comfortable to you, or are the options restricted?

Pins?
How many do you need? Inputs and outputs? Analog or digital?

Cost?
How much do you want to pay for the board? Can you use it again for other projects?

Processing power?
Will it handle all of the things you want it to do quickly and efficiently?

Communications?
How will you send and receive data? Bluetooth? Wi-Fi? Zigbee? USB? UART? I2C?

Resources?
Is there a thriving community online? Does the manufacturer offer support?

Not only will it be entertaining, you will learn valuable lessons

Written by Paloma Fautley

PALOMA FAUTLEY
is a builder of robots, both terrible and solidly mediocre. She loves learning new skills and helping others learn them as well.

STEP 2: Lower the Bar

Don't expect to do something amazing on your first try. Make the goal to learn something new. You also may want to change your current definition of robot — they have a wide range of shapes, sizes, and purposes.

STEP 5: Make It Do Stuff!

Your first question is probably "Why? I want to build awesome robots, not stupid ones and I want to do it the first time and never make mistakes."

Well, terrible robots are hilarious. Look at the bots from Simone Giertz or HeboCon and tell me that's not great. And, building terrible robots helps you build awesome robots. The more you learn along the way, the more you can apply to your next build. ⊘

Outputs:

Robot's brain sends a command to do something

» Run at a specific voltage
» You may need an additional power source to run it

Spinning a motor

There are a ton of different types of motors out there, but to learn more about DC, stepper, and pulse width modulation (PWM) motors, visit this project online at makezine.com/go/build-crappy-robots.

» Stepper
» Servo
» DC
» DC with H-bridge
» Motor controller

Turning on a light

Individual or strip LEDs

» What (forward) voltage is needed to turn on the LED?
» What current is needed to run all the LEDs at once?
» What signal is needed to control the LEDs?

LED/LCD panel Multiplexing

» Turns on only one row at a time very quickly
» Looks like all of the LEDs are on at once to your brain
» Saves a lot of power

Playing a sound

» **Speaker** has a wide range of tones
» **Buzzer** usually is one note

Small-Scale
Solar Power

Transform a 25-watt semi-flexible panel
into a practical battery charger

Written and photographed by Forrest M. Mims III

TIME REQUIRED:
1–2 Hours
COST:
$100–$150

MATERIALS
» Semi-flexible solar panel, 20-watt to 25-watt
» Lithium battery pack with 15V–20V solar input charging port
» #12 wire (red and black)
» Heat-shrink tubing
» Appropriate battery connector

TOOLS
» Wire cutters
» Soldering iron
» Heat-shrink tubing

The author's DIY solar panel from 1975.

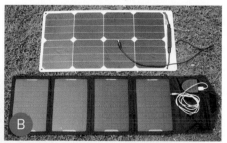

Foldable and semi-flexible solar panels.

This solar panel became almost hot enough to fry an egg.

A 100-watt solar panel scalded the grass where it was placed for a day.

Sunlight may be free, but transforming it into usable electricity is not. I learned this back in 1965 when I bought some surplus silicon solar cells from dealer Herbach and Rademan (now H&R Company, Inc.). Those 0.75"×0.75" cells cost around $2.49 each or $19.30 today according to the CPI Inflation Calculator.

In 1975, I used nine of those solar cells to make a charger for two AA penlight cells that powered my flashlight during cross-country bicycle trips. I placed the cells on a 3"×6", ⅛"-thick acrylic sheet after they were soldered together, covered them with Silastic adhesive and placed a thin sheet of clear plastic over the Silastic. I added a blocking diode between the positive output and the positive battery holder terminal to keep the batteries from discharging through the solar cells at night. This provided a waterproof solar charger (Figure A) that works as well today as it did 42 years ago.

ADVANTAGES OF PORTABLE SOLAR POWER
Today's silicon solar cells are more efficient and cheaper than those I bought in 1965. They are also available in a wide variety of preassembled, waterproof, and even foldable and semi-flexible (Figure B) arrays with built-in blocking diodes. Semi-flexible arrays are much lighter than those installed in metal and glass frames. Non-silicon arrays are also available, but they don't provide the efficiency and long life of silicon.

Portable solar power is ideal for charging phones and flashlight batteries during bicycle trips, hikes, and camping. It's especially handy during extended power failures. But solar power also has drawbacks.

THE COST OF PORTABLE SOLAR POWER
Cost is the major drawback of portable solar power. As these words are being typed, my iPhone SE is consuming 5.3 watts (0.0053 kW) while being charged through a Kill A Watt power monitor. This phone requires 2.5 hours to be fully charged from empty. Assuming the power consumption during charging is constant (mine declines

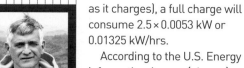

FORREST M. MIMS III (forrestmims. org) an amateur scientist and Rolex Award winner, was named by Discover magazine as one of the "50 Best Brains in Science." His books have sold more than 7 million copies.

as it charges), a full charge will consume 2.5 × 0.0053 kW or 0.01325 kW/hrs.

According to the U.S. Energy Information Agency (eia.gov), in May 2017 the average cost of electricity in the 50 states was 13.02 cents per kilowatt-hour. At this price, fully charging my phone every day for a year costs only 62.97 cents.

This test is the first time this phone has been charged by line current since spring 2016, when I began charging it from a lithium power pack charged by a 25-watt, semi-flexible solar panel with a combined cost of more than $150. I consider this a reasonable expense for hikes, camping, and emergencies.

OPERATIONAL PROBLEMS
Solar power works only when the sun is shining, even when thin cirrus cloak the sky. But thicker clouds significantly reduce maximum energy production from a panel. Another consideration is that solar panels should be pointed toward the sun for best results. This is a good excuse for me to get away from my writing desk several times a day to adjust the position of my panels. Another issue is that solar panels can become very hot, especially during summer days. While I've not yet been able to fry an egg on a solar panel, I've come close (Figure C). Lying a panel on grass on a hot day will bake the plants (Figure D). The heat problem is why rechargeable battery packs with built-in solar chargers may not be a good idea, at least during summer.

SOLAR CHARGING CAMERAS, MOBILE PHONES AND TABLETS
Compact panels like the foldable array in Figure B are equipped with a USB port for directly charging devices that have a USB power port. While these panels can be easily carried in a backpack, they must be used with caution to avoid overheating the device they are charging. Advertising photos sometimes show these panels adjacent to or even behind a phone, which means the phone's internal battery pack might overheat. Instead, it's best to place the battery or device being charged away from

A thick layer of insulation will protect a battery pack from the heat of a solar panel.

Connecting a solar panel's wires to the power plug of a storage battery.

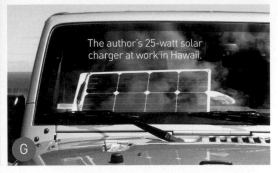

The author's 25-watt solar charger at work in Hawaii.

both the panel and direct sunlight.

Indirect charging is when a solar panel charges a battery pack that later charges various devices. For the past year, an ATOTO Ultra UPS Power Source has resided on the left side of my desk, where it is now charging my phone and a tablet computer. Every 10 days it is recharged outdoors with a 100-watt solar panel. While the ATOTO is apparently no longer available, the Chafon CF-UPS018 346WH Portable Power Supply is a very similar unit with better capabilities. I also have several compact lithium power packs with built-in charge controllers that can be charged by 15- to 20-volt solar panels. While some of these are no longer available, many USB-compatible battery packs are available that can be charged by solar panels with a USB output. Figure E shows my 25-watt panel charging a power pack during a hot summer day when the temperature was 98°F. When the power pack reached a temperature of 115°, I moved it indoors.

A DIY PORTABLE SOLAR CHARGER

When traveling, I carry a foldable solar panel suitable for charging devices and battery packs with a USB port (Figure B). On long trips and at my home office, I use 25- and 100-watt semi-flexible panels designed to charge high-capacity battery packs at 15 to 20 volts. The 25-watt panel has been modified to charge a RAVPower Model RP-PB14 lithium battery pack, a ChargeTech

(expensive) and an ANKER Mobile Power 79AN20L (discontinued). Like most high power panels, the 25-watt panel came with positive and negative cables terminated with weatherproof MC4 connectors designed to interconnect two or more panels in series or in parallel. Here's how I transformed the 25-watt semi-flexible solar panel into a practical battery charger (Figure F):

» Clip off both power leads three inches from the plastic terminal at the top of the panel and remove 0.5" insulation from the end of each wire.

» Use a multimeter to determine the polarity of the leads while the panel is exposed to weak sunlight or indoor incandescent lighting. Mark the upper corners of the panel with the respective polarities (+ and -).

» Slide 1" lengths of heat-shrink tubing over each of the panel wires.

» Remove 0.5" insulation from both ends of a pair of 16" red and black #12 stranded wires and insert the wires through the two uppermost eyelets from the back side of the top of the panel (red = + eyelet and black = - eyelet).

» Observe polarity and solder the new wires to the two solar panel wires.

» Slide heat-shrink tubing over the soldered connections and apply heat.

» Solder the free ends of the red and black wires to a power plug compatible with the power input socket on a compact battery pack designed for solar charging at 15 volts–20 volts (not USB). Or, to guarantee compatibility, clip a 16" section of the power brick output cable supplied with the battery pack and solder it to the red and black solar panel leads.

⚡ CAUTION: Observe polarity.

You can easily modify these steps. For example, you can drill out the two eyelets so the panel wires can fit through them.

This 25-watt system was put to the test when I calibrated NOAA's world standard ozone layer instrument at Hawaii's Mauna Loa Observatory during summer 2016. The panel was packed in the bottom of a checked bag along with 45 pounds of electronic gear. It was used every day to top off a battery pack. On days when I drove down to the coast for a shower and groceries, the panel was placed inside the windshield of the passenger side of my rental Jeep (Figure G). After 64 days in Hawaii, the panel was shipped home, where it arrived with no damage and has been used regularly ever since. To be sure you have a similarly good experience, go online and carefully review portable solar power options. ◐

Twirl-A-Squirrel

Mod your bird feeder to (gently) evict furry thieves via remote control

LARRY COTTON has finally given up doing anything earthshaking. He loves electronics, music and instruments, computers, birds, his dog, and wife — not necessarily in that order.

Written and photographed by Larry Cotton

NOTE: No squirrels were harmed with this device!

TIME REQUIRED:
1–2 Weekends

COST:
$70

MATERIALS

- » **Toy airplane** — to extract receiver circuit board, antenna, remote control, battery housing and screws Amazon #B00PAHKF26, amazon.com
- » **Alkaline Cordless screwdriver** Amazon #B004HY3APW
- » **Alkaline batteries, AA (6)**
- » **Wire, 22 gauge (2')**
- » **Alligator clip**
- » **Machine screw, 8-32×1⅝" with washer and nut**
- » **Brass brazing rod, 3/32" dia. (4")** or wire coat hanger
- » **Coat hanger (18")**
- » **Spring, 4½" long, 15/32" OD, .041" wire**
- » **Small spring, electrical tape, etc.**
- » **Aluminum, 1/16" thick (4"×4")**
- » **Acrylic sheet, ⅛" (8"×10")** for arm, cover top, and spacer. Scraps are usually available from window glass dealers.
- » **Sheet metal screws, #6×⅜" (12)**
- » **Sheet metal screws, #6×¼" (2)**
- » **Sheet metal screws, #8×⅞" (2)**
- » **Microswitch, SPST, normally off**
- » **Switch screws and nuts, 6-32 (2)**
- » **Wooden dowel, ⅜" (11/16")**
- » **PVC pipe, 4" 1120 SDR (10')** need 4¾" for cover. A similar plastic food (12oz Ovaltine, 40oz Jif peanut butter, etc.) or 2-liter soft-drink container would work. Or you can just put a one-gallon storage bag over everything.
- » **Rigid plastic, 1/16" or acrylic** for cover top if not using food containers or plastic bag
- » **Creatology foam sheet (9"×12")**
- » **Foam**
- » **Wood, 1×4 (3')**
- » **Pivot bolt, ¼-20, washers (3) and nuts (2)**
- » **PVC tubing ½" ID (nominal) (14")**
- » **Cable tie, 4"(2)**
- » **Bird feeder and sunflower seeds**

TOOLS

- » **Standard hand tools**
- » **Jigsaw and fine blades**
- » **Band saw (preferred)**
- » **Hacksaw**
- » **Phillips screwdriver, #0** for airplane screws
- » **Wire cutting pliers**
- » **Sandpaper, 120 and 320 grit**
- » **Hot-glue gun and glue**
- » **Disk sander, hand or bench**
- » **Portable drill**
- » **Drill press (preferred)**
- » **Drill bits, 1/16"–⅜" in 64ths**
- » **Soldering gun, small**
- » **Solder, thin**
- » **Ruler, tape and straight**
- » **Digital calipers (preferred)**
- » **Sharp pencil**
- » **Files, flat and rat-tail**
- » **Masking tape**
- » **Lubricating oil**
- » **Heat-shrink tubing**
- » **Heat gun (optional)**

L et's be clear at the outset: this gadget throws hungry squirrels from a bird feeder. Where we live, squirrels are a nuisance. Industrious, intelligent animals, yes, but still a nuisance. So we spend lots of time thinking of clever ways to outsmart them, but inevitably they're back hogging our bird feeders within a few minutes or hours.

One of the most effective means of at least temporarily dissuading the creatures is to throw them off said bird feeder, and you can buy a feeder that does just that. When a squirrel sits or hangs on it, its weight turns on a motor, which immediately throws it off. Find it here drollyankees.com/product/yankee-flipper-bird-feeder, at only $165 — yikes!

Unfortunately with this, if you're not around when a squirrel tries to gobble the goodies, you miss all the fun. So I decided to throw the creatures off by remote control. Sure, they still get to eat — sometimes a lot —but that's more than offset by the fun of interrupting a greedy, complacent, seed-stealing squirrel by pressing a button.

So the squirrel's mad now and doesn't come back? Hardly; he's very curious and tries to figure out what just happened. Or maybe he enjoys being suddenly ejected from the dining table like an unruly child. Inevitably, he tries it again, and ... well, see the previous paragraph.

Holy Stone! An RC transmitter and receiver for $11! Yes, our remote control radio comes from Holy Stone, an Amazon supplier, cleverly disguised as a toy airplane (Figure Ⓐ). Amazon also sells a squirrel-twirling motor for $9 cleverly disguised as a Black & Decker screwdriver (Figure Ⓑ).

1. HARVEST USEFUL PARTS FROM TOY

Remove the plane's top housing with a #0 Phillips screwdriver. I didn't use its motor (Figure Ⓒ, in the squarish yellow housing), but if you want to save it for future projects, carefully lift both halves out together. Otherwise the gears will fall out and it's a challenge to put them back in again; trust me. I wrapped a 4" cable tie around the two motor parts and stashed it with my other surplus DC motors.

Remove the rest of the parts and don't lose the screws. Save the receiver circuit board, the antenna and, for now, the plane's lower body (Figure Ⓓ). Use the gaudy little

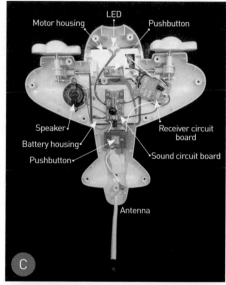

Motor housing — LED — Pushbutton
Speaker — Receiver circuit board
Battery housing
Pushbutton — Sound circuit board
Antenna

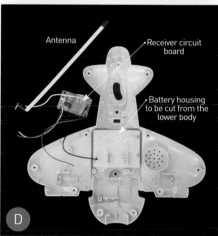

Antenna — Receiver circuit board
Battery housing to be cut from the lower body

NOTE: There's a chance that wire colors and other details mentioned here won't match those in your plane. For instance, look at the wires in Figure Ⓓ. If, say, the red wire in the corner of your receiver board is actually green, substitute "green wire" everywhere you see "red wire" in this text. Do the same for any other discrepancies.

transmitter with no modifications. Clip unneeded bits from the receiver board: the switch, LED, speaker, the sound circuit board, and the pushbuttons.

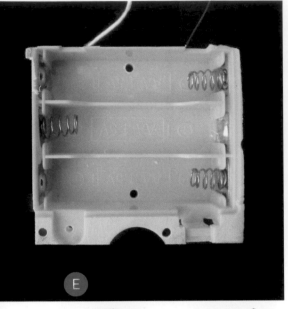

E

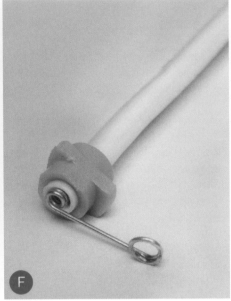

F

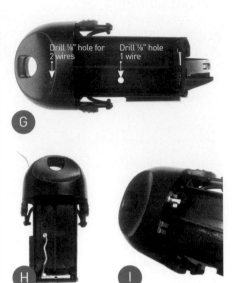

G

H

I

J

K

L

M

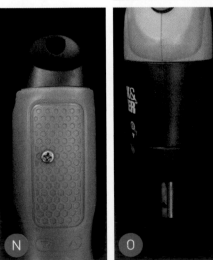

N

O

2. PREPARE THE BATTERY HOUSING AND ANTENNA

Cut the battery housing from the plane's lower body; don't damage its 4 walls or 2 wires. Lightly sand the sawed edges. Overall dimensions of the housing when you're through should be about 2⅝"×2". Drill two ³⁄₃₂" screw holes as far apart as possible in the battery area, halfway between the ends of the batteries, for mounting (Figure E). Be sure to allow for the diameter of the screw heads. Recycle the rest of the lower body and any unused parts.

Hot-glue all the spots where wires exit the receiver and battery housing. Since they're somewhat fragile, make notes of what color wires go where in case they break. Set these parts aside. Sand-flatten one or more sides of the antenna's blue part so that it can mount almost flush to a surface (Figure F).

3. MOD THE SCREWDRIVER BATTERY COMPARTMENT

Make sure your screwdriver runs by installing 4 AA batteries in its battery compartment, then alternately pressing the forward/reverse buttons on its side. Remove and install the battery compartment a few more times. Still runs? Great! Remove the testing batteries.

The airplane circuitry uses 4.5V from 3 AA cells, but to give the squirrels a good spin, I added another AA battery in series with the screwdriver motor — not to the rest of the circuit. It's housed solo in the screwdriver

battery compartment, which normally takes 4 AA batteries.

In the end of the battery compartment, to one side of its centerline, drill a ⅛" hole to allow two wires to protrude. Note the position of the terminals that fit inside the screwdriver. Inside, drill another hole to clear one wire (Figure G). Locations aren't critical.

Use approximately 22-gauge wire to wire the battery compartment. I was able to solder only to the metal pieces that connect with the socket in the screwdriver (Figure H), so I added a small alligator clip to connect to a terminal at the other end. Bend the terminal up slightly (Figure I) to make it easy to grab with the alligator clip.

My purple wire, soldered as shown in Figure J, will go to the blue wire on the receiver and my yellow wire, soldered as shown in Figure K, will go to the white wire on the receiver. Run both wires out of the first hole you drilled. The additional hole visible here is for my insurance screw in the next step.

To suspend the screwdriver from the feeder arm, saw a slot in the center of the battery compartment ⅛" wide (Figure L). Don't cut more than ⅞" deep or you could damage the metal battery terminals.

4. ADD AN INSURANCE SCREW

Although it's never happened to me, a vigorous pull on the screwdriver might separate the orange screwdriver housing (thus the bird feeder, greedy squirrels, etc.)

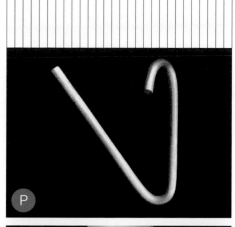

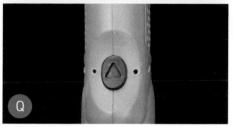

(R) **Arm** ⅛" clear acrylic

**Drilled holes (shown in red)
identification and diameters:**
A – pivot bolt (¼")
B1 – vertical travel limit screw (⅜")
B2 – screwdriver attachment dowel (⅜")
C1 – lower spring attachment (¹⁄₁₆")
C2 – battery housing attachment screws (¹⁄₁₆")
C3 – receiver board attachment screws (¹⁄₁₆")
D1 – switch screws (³⁄₃₂")
D2 – antenna cable tie (³⁄₃₂")
D3 – cover screws (³⁄₃₂")

from its black battery compartment. To prevent this, I ran a long machine screw through the screwdriver from one side to the other, passing through the battery compartment.

Put one AA battery into its compartment. Decide where you want to drill a hole, which will determine where you'll drill the hole in the screwdriver housing. The drill bit must miss the battery, the compartment's ribs, and all of your new wiring. Mark the spot, but don't drill it yet!

Measure the distance from the interface of the two housings and the offset from centerline to your marked spot, then mark the same dimensions on the screwdriver housing. Push the battery compartment into the screwdriver housing until it locks. Holding both parts tightly together on a flat surface, drill an ¹¹⁄₆₄" hole all the way through (Figure (M)). A drill press helps with this operation, but it's not critical.

As a clearance check, push an 8-32 × 1⅝" machine screw from one side of the screwdriver through the other. Figure (N) shows the location of my insurance screw; yours may differ somewhat. If OK, remove the screw and screwdriver from the battery compartment for now. Remove the battery.

5. MOD THE CHUCK AND FORWARD/REVERSE BUTTONS
File or disk-sand a small flat spot near the end of the screwdriver's chuck, then drill — preferably with a drill press — a ³⁄₃₂" dia. hole into the flat (Figure (O)). The

chuck has been hardened, which will resist penetration. Use a new bit if possible, and keep its tip lubricated with a few drops of oil. Drill until you have a hole in one side of the chuck.

Make a hook out of a coat hanger, or better, a ³⁄₃₂" piece of brass brazing rod. Bend it approximately as shown in Figure (P) and twist it so the small and large bends are not in the same plane (30°–45° apart is fine). Loop the smaller hook into the hole you just drilled and thread a small spring over the hook to hold it in position. Electrical tape will do if you can't find a spring to fit.

The screwdriver won't run unless one of the buttons on its sides is held in, so cut a button-holding strap about ¼" × 1-³⁄₁₆" from ¹⁄₁₆" thick aluminum. It's a small part, so be careful; keep your hands away from the blade! Drill two ⅛" holes in it. Use it as a template to carefully drill ³⁄₃₂" dia. matching holes in the screwdriver body on both sides of either button (Figure (Q)).

⚡ CAUTION: Wrap a piece of tape around the drill bit ⅛" from its point to keep from drilling too deep.

Attach the strap to the screwdriver with two #6×¼" sheet metal screws just until the button is held in. Don't overtighten the screws.

6. MAKE THE ARM
Cut out the arm from a new piece of acrylic with protective sheets on both sides. It will hold the receiver, antenna, battery

housing, and a new switch. It will also suspend the screwdriver, which in turn suspends a bird feeder.

Drill all the holes shown in red, except C1 (Figure (R)). Be careful drilling the holes for the battery housing, receiver board, and switch; those holes must match corresponding holes in the parts. For those parts, you could temporarily fasten them in position with double-sided tape, then trace the mounting holes onto the acrylic protective sheet. That will ensure the mounting screws line up with their respective holes in the arm.

IMPORTANT: Hole C1 is a tension spring attachment point, which takes a fair amount of (upward) force. I've positioned that hole's center ³⁄₃₂" from the arm edge; don't drill any closer to the edge than that!

7. MOUNT PARTS ON THE ARM
Mount the receiver board and the airplane's battery housing using four of the screws you saved from plane disassembly. Don't overtighten the screws.

Mount the switch with two machine screws and nuts. The switch button must protrude past the edge of the arm. Bend the switch terminal(s) if necessary. Solder the red wire from the receiver, and the yellow wire from the battery housing, to the switch. Don't break the black wire from the negative terminal of the battery housing to the receiver. Mount the antenna with a 4" cable tie and connect its short, stiff wire to

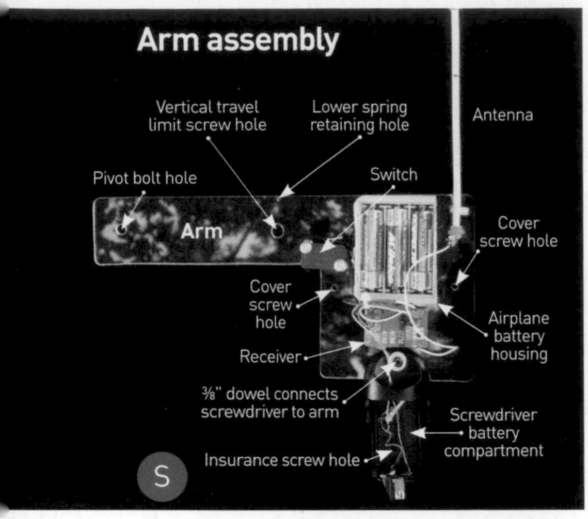
Arm assembly
Vertical travel limit screw hole
Lower spring retaining hole
Antenna
Pivot bolt hole
Switch
Arm
Cover screw hole
Cover screw hole
Airplane battery housing
Receiver
3/8" dowel connects screwdriver to arm
Screwdriver battery compartment
Insurance screw hole
S

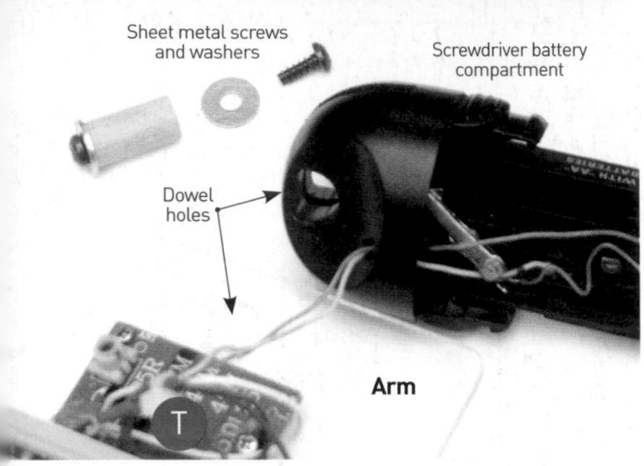
Sheet metal screws and washers
Screwdriver battery compartment
Dowel holes
Arm
T

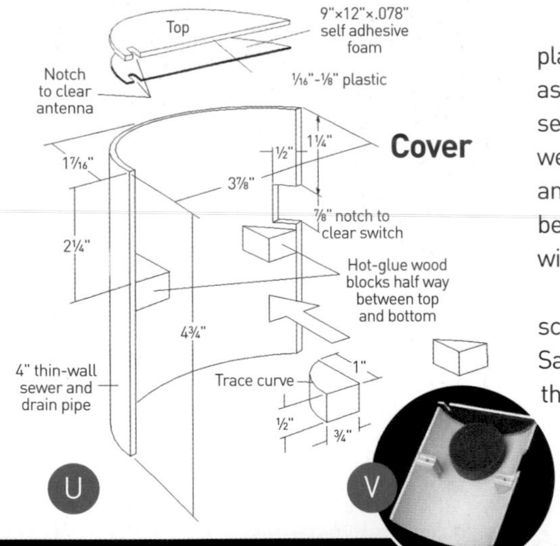
Top
9"×12"×.078" self adhesive foam
Notch to clear antenna
1/16"-1/8" plastic
Cover
1 7/16"
1/2"
1 1/4"
3 7/8"
7/8" notch to clear switch
2 1/4"
Hot-glue wood blocks half way between top and bottom
4 3/4"
4" thin-wall sewer and drain pipe
Trace curve
1"
1/2"
3/4"
U
V

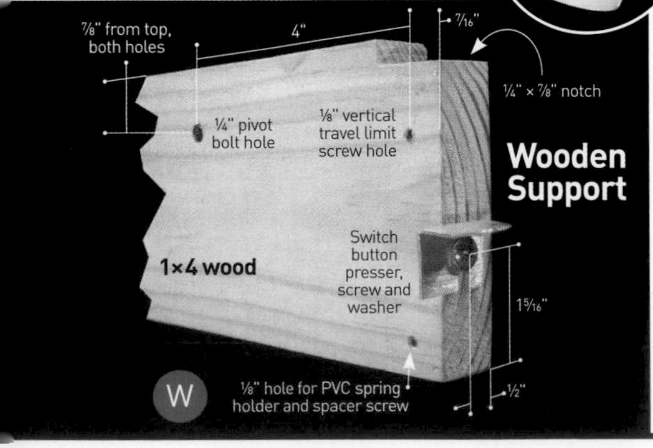
7/8" from top, both holes
4"
7/16"
1/4" × 7/8" notch
1/4" pivot bolt hole
1/8" vertical travel limit screw hole
Wooden Support
Switch button presser, screw and washer
1×4 wood
1 5/16"
1/8" hole for PVC spring holder and spacer screw
1/2"
W

the receiver's wire with either a sheet metal screw or a small machine screw and nut (Figure S).

Drill two 1/8" pilot holes into the ends of an 11/16" long × 3/8" diameter dowel; it'll be used to lock the screwdriver to the arm (Figure T). Slip the screwdriver's battery compartment over the arm so that its battery will be on the opposite side of the arm from the 3-battery holder. Insert the dowel (some sanding may be necessary), then add washers and screws.

Solder your two new wires from the battery compartment, as shown in Figures J and K, to the blue and white leads on the receiver board. Reversing this polarity will fry stuff, so be careful. Use heat-shrink tubing on the soldered joints.

8. MAKE THE COVER

I made the curved shell from a piece of 4" inside diameter, thin-wall PVC pipe I had on hand, but a similar plastic food (12oz Ovaltine, 40oz. Jif peanut butter, etc.) or 2-liter soft-drink container would work. Cut a notch in one side to clear the switch. Or you can skip the cover altogether if you don't mind putting a one-gallon storage bag over everything — the squirrels won't care.

I made the cover top from 1/16" rigid plastic (excess from arm acrylic will work, as well) and attached a piece of Creatology self-adhesive foam under it to form a weatherproof seal, then hot-glued the cover and top together (Figure U). The top won't be necessary if you use a food container with a relatively flat bottom.

Make two cover-mounting blocks from scrap wood by tracing your cover's contour. Sand the mating surfaces, then hot glue them into place. Hot-glue a small piece of foam inside the cover to hold the 3 batteries in (Figure V).

9. MAKE THE WOODEN SUPPORT (AKA SQUIRREL CROSSWALK)

Cut it from standard 1×4 (3/4"×3 1/2" actual) wood. Its length will depend on your keeping the bird feeder about 2' away from the tree the support is mounted on.

Notch one end to clear the bottom coil of the tension spring. Drill a hole for the pivot bolt and 3 pilot holes for mounting the tension spring holder and the switch button presser (Figure W).

10. MAKE THE TENSION SPRING HOLDER, SPACER AND SWITCH BUTTON PRESSER

Cut the 1/2" (nominal inside diameter) PVC pipe 14" long. Later you can shorten it after compensating for the weight of your bird feeder. My feeder weighs 1 1/2lbs and the pipe length is 12".

Cut the L-shape opening; it fits over the wooden support.

Drill two 11/64" holes in the cutout end to clear mounting screws. Drill several 1/16" holes near the top, through both sides, for a nail, which will hold the upper end of the tension spring (Figure X). The location of this nail controls upward spring force, counteracting the weights of the arm assembly, the bird feeder, and — you guessed it — a squirrel.

Make the spacer from excess arm acrylic (Figure Y). Carefully cut out the switch button presser in its flat shape with a band saw or jigsaw in a vise. Cut its slot with a hacksaw and/or a file. Bend 90° as shown (Figure Z).

11. FINAL ASSEMBLY AND TESTING

Firmly clamp the wooden support horizontally in a vise. Attach the arm assembly to the wooden support (Figure AA) with a 1/4"-20 bolt, washers between parts and 2 nuts (the outer nut locks the inner one). The 3 1/2" side of the support must be vertical, and the arm must pivot freely with no binding.

Attach one end of the tension spring to hole C1 (refer to Figure R) in the arm. Drop the tension spring holder over the spring and attach the holder to the wooden support arm with two #8×7/8" sheet metal screws. Use the acrylic spacer on the bottom screw (see Figure AA). The top screw also serves as a vertical travel limit for the arm. Leave both screws just loose enough for the arm to move freely.

Bend a small hook at the end of an 18" piece of straightened coat hanger and reach into the top of the spring holder until you can grab and stretch the spring upward. While pulling up, insert a nail to hold the spring at the top of the tubing. Remove the coat hanger.

Holding the cover in position, trace the 2 holes you've already drilled in the arm (D3 in Figure R) onto the mounting blocks. Drill two 1/16" pilot holes in the blocks. Add 3 AA

GIVE A GIFT.
6 ISSUES ONLY $39.99.

Make:

GIFT FROM

NAME _____ (PLEASE PRINT)

ADDRESS/APT. _____

CITY/STATE/ZIP _____

COUNTRY _____

EMAIL ADDRESS (required for order confirmation) _____

☐ Please send me my own subscription of Make: 6 Issues for $39.99.

GIFT TO

NAME _____ (PLEASE PRINT)

ADDRESS/APT. _____

CITY/STATE/ZIP _____

COUNTRY _____

EMAIL ADDRESS (required for access to digital edition) _____

We'll send a card announcing your gift. Make: is published bimonthly. Allow 4-6 weeks for delivery of the first issue. Your recipient can also choose to receive the digital edition at no extra cost. For Canada, add $9 per subscription. For orders outside the U.S. and Canada, add $15. Payable in U.S. funds only.

481GS1

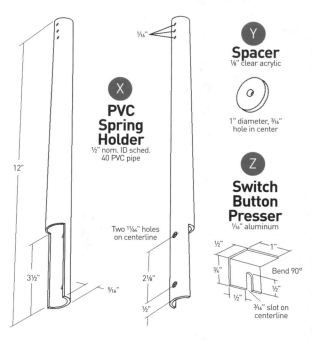

X

PVC Spring Holder
½" nom. ID sched. 40 PVC pipe

12"

3½"

⁵⁄₁₆"

2⅛"

½"

Two ¹¹⁄₆₄" holes on centerline

¹⁄₁₆"

Y

Spacer
⅛" clear acrylic

1" diameter, ³⁄₁₆" hole in center

Z

Switch Button Presser
¹⁄₁₆" aluminum

½"   1"

¾"   Bend 90°

½"

½"   ½"

³⁄₁₆" slot on centerline

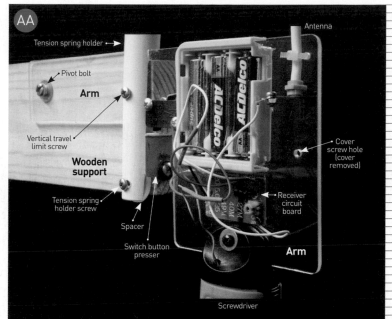

AA

Tension spring holder

Antenna

Pivot bolt

Arm

Vertical travel limit screw

Wooden support

Tension spring holder screw

Spacer

Switch button presser

Cover screw hole (cover removed)

Receiver circuit board

Arm

Screwdriver

batteries (watch polarity), then mount the cover with a couple of #8×½" sheet metal screws. Don't seal it; you'll need to get to the batteries eventually.

Mount the switch button presser with a #6×⅜" sheet metal screw per Figure **W**. Leave a small gap between the bottom of the button and top of the presser. Put a battery in the screwdriver battery compartment observing polarity, then push the screwdriver body over it. Add the 8-32 × 1⅝"-long insurance screw, a washer and a nut. Tighten securely.

As already mentioned, the arm must freely pivot a few degrees on the pivot bolt — enough to push and release the switch button. It controls power to the receiver. (The screwdriver won't run until the transmitter button has been pressed.)

Put 2 AA batteries in the transmitter, push down on the arm and press a button on the transmitter. One button should turn the screwdriver fast in one direction; the other button should turn it slowly in the other direction.

Fill your bird feeder (sunflower seeds are light and tasty) and hang it on the bird feeder hook. Add anything that weighs about 8 ounces (average weight of a hungry squirrel) to the bird feeder tray.

With 8oz on the tray, the button must be barely pressed. With no weight on the tray, the button must be barely not pressed. Adjust this by using the hooked coat hanger or rod to move the top loop of the tension

BB **Adjusting the Tension Spring**

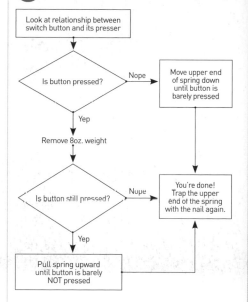

Look at relationship between switch button and its presser

Is button pressed? — Nope → Move upper end of spring down until button is barely pressed

Yep

Remove 8oz. weight

Is button still pressed? — Nope → You're done! Trap the upper end of the spring with the nail again.

Yep

Pull spring upward until button is barely NOT pressed

spring up or down. You may have to drill another hole or two in the PVC tubing for a new nail position. Figure **BB** might help. Once again, the arm must freely pivot. Remove the 8oz weight.

The fun begins! Mount your squirrel slinger close to a tree and not too far off the ground. Disappear and chill for awhile to give the squirrels time to find it. Let a squirrel climb onto the feeder, then press a button on the transmitter. One of the buttons will turn the feeder faster than the other.

Hilarity ensues. ●

TROUBLESHOOTING

1. The usual suspects: cold solder joints, wiring, mis-polarized batteries, etc. Check that one (either) side button on screwdriver is firmly pressed in.

2. Rarely, I've encountered bad contact between the battery compartment's plug and the screwdriver's socket. The plug's terminals can be bent outward by carefully inserting a No. 11 X-Acto blade and twisting.

3. There may be too much friction between the acrylic arm and the wooden support. Keep the support parallel to the ground, and its 3½" face vertical. The tension spring holder should be mounted loosely. Add enough washers to keep the parts clear of each other. Use Teflon spray as a last resort.

4. Re-read the paragraph regarding adjusting the tension spring with food in the tray under Final Assembly and Testing. Overall weight of the rig decreases as squirrels and birds consume food, and you may need to add weight to the feeder to compensate. Obviously, refilling is the best choice, but in a pinch I put a couple of the larger sockets from socket-wrench sets into the tray. The birds and squirrels ignore them.

Check out makezine.com/go/twirl-a-squirrel-bird-feeder for video footage of the Twirl-A-Squirrel in action.

Boost Your
Bubbles

Combine helium and air to float clouds of soapy water

Written by Marcos Arias and Harrison Fuller

MARCOS ARIAS
is a teacher working
in North Hollywood
at Oakwood School,
where he teaches 7th–
12th grade classes
focused around STEAM
(Science, Technology,
Engineering, Art and
Math).

HARRISON FULLER
is going into his senior
year at Oakwood and
plans on majoring in
engineering after he
graduates. He started
working on the Bubble
Printer in 10th grade as
an independent study.
He also has a deep
passion for military
history and making
things that explode.

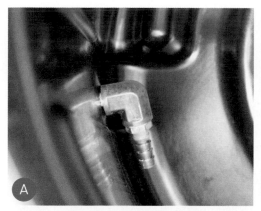

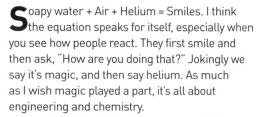

Jordan Weiner

TIME REQUIRED:
A Weekend
COST:
$380–$550

MATERIALS
» Trash can, 20 gallon
» Barb, ¼" male NPT × ⅜"
» Barb, ¼" male NPT × ¼" (3)
» Elbow, 90°, ¼" male NPT × ¼" female
» Elbow, 90°, ¼" female × ¼" female
» Vinyl hose, ⅜" inner diameter (6')
» Vinyl hose, ¼" inner diameter (2'–3')
» Galvanized wire, ⅛", 6 gauge (5'6")
» Bolt, ½"
» Hose clamps, ¼" to ½" (6)
» Welding torch hose
» NPT plug, ¼" female
» NPT plug, ¼" male (2)
» NPT coupler, ¼" female
» Welding torch tip
» Co2 pressure regulator
» On/Off ball valve, NPT male × NPT female ¼"
» Plywood, ¾", 20"×20", 6"×14", and 3"×4.5"
» Wood, 2×2" (8')
» Pipe strap, ½" (2)
» Welding torch
» Screws, 2" (17)
» Screws, 1.25" (6)
» PVC pipe, ¾" (3')
» Ajax or Dawn dish soap
» Corn syrup
» Distilled water, warm (1 gallon)
» Small air compressor
» Helium tank (55 cu ft)
» NeverWet 2-step kit (optional)

TOOLS
» Tape measure
» Drill
» Drill bits, ½" and 1.4mm
» Sharpie
» PTFE tape
» Crescent wrench
» Wrenches, 9/16", 11/16", 5/8", ¾"
» Wire cutters, large and regular
» Flathead screwdrivers
» Clamps (2)
» Hacksaw
» Glue gun and glue
» Cyanoacrylate (CA) glue aka super glue
» Duct tape
» Measuring cups and spoons
» Small container
» Spray bottle with soapy water
» Sandpaper (optional)

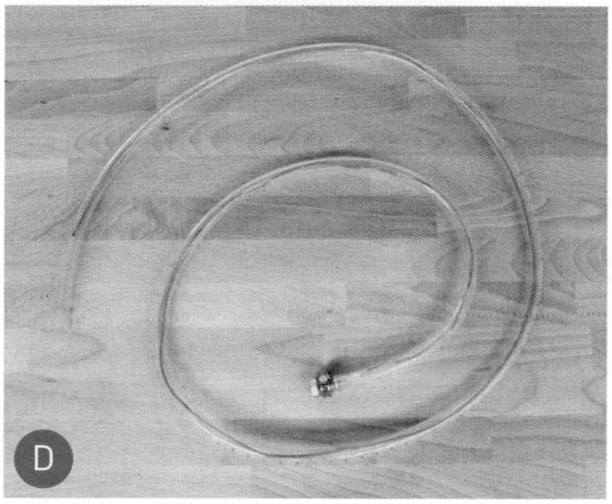

Soapy water + Air + Helium = Smiles. I think the equation speaks for itself, especially when you see how people react. They first smile and then ask, "How are you doing that?" Jokingly we say it's magic, and then say helium. As much as I wish magic played a part, it's all about engineering and chemistry.

My motto in the STEAM Department is "I hear I forget, I see I remember, I MAKE I understand." I love to make and so do my students. So I asked one of our students if he wanted to do an independent study project to build a machine that made bubbles float. That was two years ago when we started and now are working on a larger version that is 4'×8'. The version here uses a trash can but you can use any container you want and adjust accordingly.

1. DRILL CONTAINER AND ATTACH MANIFOLD FITTINGS
Drill a ½" size hole, 2.5" up from the bottom of the trash can. Place PTFE tape on the male fittings of the ¼" male × ⅜" barb and the ¼" male × ¼" female 90° elbow. Connect the barb to the ¼" female × ¼" female 90° elbow using the 9/16" and 11/16" wrenches and push it through the hole from the inside of the trash can (Figure Ⓐ).

Take the ¼" male × ¼" female 90° elbow and install it from the outside of the can and tighten

(Figure Ⓑ) — you do not need to worry about water leaking through the hole because the waterline will be below it.

2. BUILD AND INSTALL MANIFOLD FOR BUBBLES
Stretch and clamp the 6' length of ⅜" vinyl hose onto a table. From the left side, mark 4" in and then mark with a Sharpie every ½", but leave 1" on the right side without holes. Drill holes at the marks with the 1.4mm bit (Figure Ⓒ) — making sure to not leave any plastic in the holes.

Take the 5'6" length of galvanized wire and run it through the tube from the right side. Quickly take the glue gun and put glue on the right side of the tube before pushing the ½" bolt in and then tighten a hose clamp, so no air leaks out. Coil the manifold like in Figure Ⓓ and make sure that the holes are facing up.

Slip a hose clamp on the open end of the tubing, push the manifold onto the ⅜" barb in the trash can, and tighten the hose clamp. Use the glue gun to temporarily make the manifold stick to the trash can, so it doesn't rise when you turn on the machine and go above the waterline. After you get the machine working and adjusted, use medium cyanoacrylate glue or something that will permanently attach it to the trash can (Figure Ⓔ).

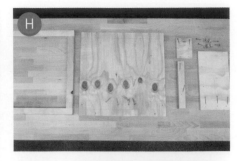

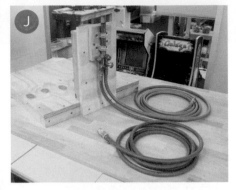

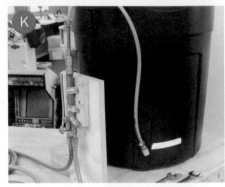

Jordan Weiner, David Sanders @ WET Design

3. BUILD AIR AND HELIUM LINES

Take large cutters and trim one side off of the green and red welding torch hose so you can attach two ¼" male NPT × ¼" barbs. Wrap both barbs with PTFE tape, then put hose clamps on the green and red hoses before pushing in the barbs on both.

Screw the ¼" female NPT plug onto the green hose and the ¼" female NPT coupler onto the red hose. Tighten with the ¹¹⁄₁₆" and ¾" wrenches, and screw the hose clamps tight on both hoses (Figure F).

Cut 2"–3" off of the welding torch tip, just enough so the ¼" vinyl hose will fit tightly onto the tip. Take the ¼" hose and push it onto the welding torch and tighten a hose clamp around it. Slide a hose clamp into the other side of the hose. Put PTFE tape on the ¼" male NPT × ¼" barb, then push it into the other end and tighten the hose clamp (Figure G).

Put PTFE tape on the male ends of the ¼" On/Off ball valve NPT male × NPT female and the ¼" female NPT plug. Attach the ¼" female NPT plug to the ¼" On/Off ball valve NPT male × NPT female. Take the pressure regulator and the piece from the previous step and attach them together using the ⅝" and ⁹⁄₁₆" wrenches.

4. BUILD WOOD STAND

Cut the 8' piece of 2×2 into two lengths of 20", two of 17", and one 12". Make the frame by screwing the two 20" and two 17" pieces together using 4 of the 2" screws, 1 on each corner (Figure H).

Put the 20"×20" piece on top of the frame and screw in place using eight of the 2" screws. Mount the 3"×4.5" piece of wood to the 6"×14" piece using 2 of the 1.25" screws. Attach the welding torch to the side piece with the 2 pipe straps and the four remaining 1.25" screws (Figure I).

Take the 6"×14" piece and screw the bottom of it into the square base on one side. Take the 12" piece of 2"×2" and two of the 2" screws and attach it to one of the sides of the 6"×14" piece. Then connect the red line to the left side and the green line to the right side (Figure J). Screw tightly by hand and finish with the crescent wrench.

5. CONNECT EVERYTHING TOGETHER

Attach the welding torch tip air line you made in step 3 to the trash can first, tightening with a wrench, and then to the welding torch tip (Figure K). Tighten with a wrench.

Secure the helium tank so that it won't fall over or put it in a tank stand. Attach the pressure regulator to the helium tank and tighten using a crescent wrench. Attach the green hose to the air compressor. Attach the red hose to the helium tank (Figure L). Make sure the ball valve is closed.

6. CHECK FOR HELIUM AND AIR LEAKS

Turn the valve to the left on the green line of the welding torch and spray soapy water anywhere there is a connection. If you see bubbles at any connection that means there is a leak and it needs to be tightened or redone with PTFE tape. Make sure to spray soapy water on the connections in the trash can, both inside and outside.

Turn off the air (green line) before checking the helium line. Make sure the ball valve is closed on the pressure regulator and the dial on the red line for the welding torch is closed. Turn open the helium tank and adjust the pressure regulator to 22 PSI (Figure M). Open the ball valve and spray soapy water on all the red hose connection points, and ball valve. When all connections have been confirmed, close the helium tank bottle and turn the ball valve off.

7. CREATE AND MARK DIAL INDICATORS

Print out templates for dials or laser cut if possible — you can find a PDF in this project online at makezine.com/go/bubble-printer. If printed, glue each to a 35×35mm square and cut a notch out in the middle. Slip the templates over the dials on the welding torch. Make sure valves are off and then mark with a Sharpie to indicate the off position (Figure N).

8. BUILD BUBBLE CUTTING WAND

To help the bubbles not stick to the PVC pipe when cutting layers of bubbles, you may sand the PVC pipe to make it rough and apply NeverWet. Put duct tape on one side of the wand to denote the handle, as visible on top of the trash can in Figure L.

9. PREP BUBBLE SOLUTION AND FILL TRASH CAN

Take 1.5 gallons of distilled water and pour it into the trash can. Make sure that the water covers the hose. The waterline should be just below the hole we cut in step 1.

Take 2 cups of distilled water warm, ½ cup of Ajax or Dawn soap, and 1 Tbsp corn syrup and mix in a small container. Let it sit until needed in the next step.

10. ADJUST HELIUM AND AIR

Turn the air on first to check the manifold and make adjustments if necessary. You need to turn the dial past 6 to push air through all of the holes and then dial back or adjust the PSI on the regulator. Once done, turn the air off at the welding torch. Adjust the PSI on the air compressor to 45 PSI on the regulator (Figure O). Tack the manifold permanently in place with medium cyanoacrylate glue.

Pour the solution that you made in step 9 into the trash can now and stir a little. Make sure to try and pour without making bubbles. Open the helium valve first, and then the ball valve. Now turn open the green and red line valves at the welding torch, green to 4 and red to 4 according to the dial indicators described in step 7.

Adjust helium pressure to somewhere between 8–10 PSI and air to 35 PSI. Wait until bubbles rise to the top (Figure P) and slice (Figure Q) using the wand made in step 8. You will need to adjust the dials to get the bubbles to float. The numbers I gave are just a starting point. Keep track of your PSI on the pressure regulator to adjust and see how much you use every 10 minutes. It should be about a 100 PSI every 10–15 minutes.

NOTE: The bubbles should be dry to the touch, if they are wet then the solution is too strong: try diluting it by adding 2 cups of water.

The solution you added should last about 30–45 minutes before you have to add more. When that happens, take 2 cups of water from the trash can and go back to step 9.

UP THE FUN FACTOR

Create templates by cutting designs out of ½" thin styrofoam or ¼" shower board (Figure R) to put over the top of the bubble printer to create cool shapes. ✱

Paste Wax

PAUL MOORE is an avid woodworker and the editor in chief of woodworkboss. com, where he aspires to help fellow woodworking enthusiasts pursue this hobby.

The when, how, and why of using paste wax for wood

Written by Paul Moore

The Big Finish

WE MAY NEVER KNOW EXACTLY WHEN HUMANS DISCOVERED WAX and thought, "Hey, this would make a great finish!" The Romans, Egyptians, and Vikings were all known to use beeswax. And we're sure that you'll find a good use for wax too, as a wood finish.

Why so? Well, most water-based and oil-varnish mix type finishes are not resistant to abrasion. Things sliding across or rubbing against them will tend to scuff or scratch them. Hard floors are the same way — that's why most are protected by a wax finish.

Your wood projects will benefit from a coat of wax also. Wax not only protects from abrasion, but also from moisture and dust.

WHAT IT IS
Most paste waxes are Carnauba wax extracted from the leaves of Carnauba palm trees (*Copernica prunifera*), which are native to northeastern Brazil. Other waxes can be used to protect wood finishes, but Carnauba wax is the most commonly sold.

WHEN TO USE IT
Wax is very seldom used as a standalone finish. Its low melting point (140°F) doesn't even protect from a hot coffee cup. Nor does it protect against alcohols. Wax is used as a final coating on top of other finishes, new and old, for bringing a luster and abrasion protection to the finish. Waxes are often used to rejuvenate older finishes, reviving the shine and filling in minor scratches, and providing a smoother finish.

Waxes have also found great favor in finishing chalk paints. Chalk paints are porous and very flat. Wax provides a final finish that adds some shine to the otherwise matte paints and fills the pores so they are less likely to stain with use.

There are quite a few brands to choose from, but don't worry that your local store might not have the "best" brand. Any of the brands available locally will do the job. Your decision will probably come down to price.

HOW TO APPLY IT
Paste wax is easy to apply. The size of the project might determine how it's done, but basically it is either brushed or wiped on,

then rubbed with a soft, dry cloth to remove excess and bring out a nice luster. The wax should be allowed to dry to a haze before buffing. Failure to do so will simply lift it off the surface. You want the solvents to evaporate before you buff the remaining wax.

The base finish must be fully cured before applying wax over it. Failure to do so can cause solvent fumes to remain trapped in the finish, which can result in bubbling and lifting. Follow the instructions on the finish's container in regard to cure times.

LIGHT VS. DARK
As wood varies in lightness and darkness, so do waxes. Manufacturers provide color variation to assist woodworkers in dealing with light and dark finishes. Scratches in a dark finish are better hidden using a darker wax. Of course you can use a light wax on dark finishes, but scratches and scuffs may show up.

It's not recommended to use a dark wax on light woods and light finishes, as the dark wax can enhance scratches instead of hiding them.

EASY CLEANUP
If you've used a disposable brush to apply the wax, just toss it. If not, use the solvent recommended by the wax manufacturer to clean the brush. If you used rags, allow them to completely dry before tossing them in the trash. Better yet, soak them in water before doing so, to prevent combustion.

TOP-NOTCH TOP COAT
Although waxes are not ideal as a finish on raw woods, they can serve in a pinch, but really shine best when used as a top coat to final finishes. That added bit of protection and luster will really make your projects stand out. Give them a try on your next project, they just might become your woodworking finish best friend! ⊘

Полина Стрелкова / Adobe Stock

Make the Switch

*Turn on or off **electronic relays** by applying or removing a specific voltage*

Written by John Wargo

JOHN WARGO is a professional software developer and author. He's a program manager at Microsoft Visual Studio App Center. You can find him on Twitter @johnwargo and online at johnwargo.com.

I'VE ALWAYS BEEN FASCINATED WITH RELAYS — they're so absolute, so binary. They're basically switches that you turn on or off by applying or removing a specific voltage. Relays aren't needed for most microcontroller-based projects, but you'll need them when you want to open/close an external circuit, or where your circuit needs to control higher voltages than your system can provide.

ANATOMY OF A RELAY

There are two common types: **mechanical relays** and **solid-state relays**. Mechanical relays use an electromagnetic coil and a physical switch; when you apply a voltage, the switch activates. Solid-state relays deliver the same result, but don't have the mechanical components; instead, they use electronic components to do the same job.

Commercially produced relays usually operate in two different modes, **Normally Open (NO)** and **Normally Closed (NC)**, depending on how you wire the connections to the module.

Figure Ⓐ shows a relay in NO mode. In this configuration, when there's no voltage applied to the control circuit (the relay at rest portion of the image), the switched circuit is disconnected and current can't flow through the connection. When you apply an appropriate voltage to the control circuit, the electromagnetic coil in the relay activates and pulls the switch closed, enabling current to flow through the switched circuit.

In NC mode (Figure Ⓑ) the opposite is true. When the relay is at rest (no voltage applied to the control circuit), the switched circuit is closed and current flows through the switched circuit. When you activate the relay by applying an appropriate voltage to the control circuit, the electromagnetic coil in the relay activates and pulls the switch open, stopping any current flowing through the switched circuit.

Two attributes drive relay switch configurations: **Pole** and **Throw**. The pole attribute describes how many individual circuits the switch controls. A **Single Pole** (SP) switch controls a single circuit. A **Double Pole** (DP) switch controls two separate circuits; there are essentially two interconnected switches, with each

Hep Svadja, John Wargo

connected to its own circuit; when you toggle the switch, both circuits are affected simultaneously.

The switch's throw attribute describes the number of circuit paths provided by the switch. A **Single Throw** (ST) has only one circuit path. With the switch thrown one way, current flows through the circuit. With it thrown the other way, the circuit is broken and no current flows. A **Double Throw** (DT) switch offers two circuit paths. With the switch thrown one way, current flows through one of the circuit paths, with it thrown the other way, current flows through the other circuit path. A DT switch can also have a center Off position between the two circuit path options. So, relays are often described as something like SPST, SPDT, DPST, DPDT, etc.

Each relay is different, but they are all basically rectangular blocks with at least 4 electrical connectors exposed: 2 for the control circuit and 2 for the switched circuit. Pay attention to the voltage and current ratings for the control circuit (normally given as ranges that tell you what voltage and current are required to activate the relay) and for the switching part of the relay (which tells you how much voltage and current the switching part of the relay can handle).

If you wire the relay into your circuit, applying voltage across the control circuit should trigger the relay, but it won't work reliably — there are latching considerations and other issues that may affect the relay's operation. There are actually several easier ways to add relays to your projects!

RELAY MODULES
Instead of buying relays and wiring them up with transistors, diodes, and resistors, many manufacturers produce relay module boards that include everything you need (although most that I've seen don't include any documentation, so you'll have to figure out your module on your own). You can buy these with anywhere from 1 to 8 or more relays attached (Figure C). Most relay modules are designed to work with Arduino

(which provides 3V) or Raspberry Pi boards (3V–5V).

MICROCONTROLLER ADD-ON BOARDS (SHIELDS, HATS, ETC.)
To make it even easier for you to add relays to your microcontroller projects, several manufacturers produce add-on boards that stack directly onto the microcontroller using the GPIO port (for Raspberry Pi) or the header pins that most other boards support. Adafruit makes a Power Relay FeatherWing (Figure D) — I've also used relay boards with Raspberry Pi, Tessel 2, and Particle Photon.

POWER PIGTAIL
The Power Pigtail, or PP, (Figure E) is a simple solution for high voltage switching scenarios that circumvents many of the safety issues that come with high voltage. It's basically a black box with a power cord and a relay with its switched connection wired across one of the conductors in an AC power plug. When you apply a specific voltage (normally 3V–5V) to the input connections on the PP (shown with two red wires connected to it) the relay triggers and AC current passes through the power cord. The PP is wired for NO operation, but you can configure the PP for NC operation as well.

VALIDATING RELAY OPERATION
For my relay work, I created a testing jig (Figure F) I could wire into each of my projects to easily determine the state of my relays. It's a series of LEDs connected to a power source (3V DC served by two AA batteries) with each LED exposed through two open leads. When I need to test a relay circuit, I wire an LED's leads to the NC connection on one of the relays, put a couple of batteries in the battery holder and test my code. When the relay triggers, the LED lights or goes dark depending on the status of the relay. ◗

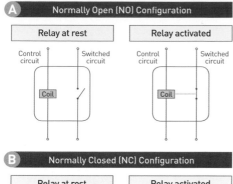
Normally Open (NO) Configuration

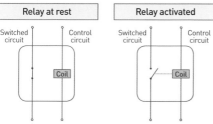
Normally Closed (NC) Configuration

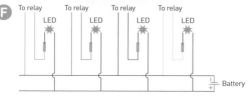

Visit **makezine.com/go/electronic-relays** for more info on relay connections, add-on boards, and triggering relays with code.

DJI SPARK FLY MORE COMBO $699 dji.com

When the DJI Spark launched, I had my doubts. I've spent time with other small, Wi-Fi controlled camera drones over the last few years and am still a little shell shocked from their wind-related performance woes and sudden signal dropouts. Sized even smaller than those ones, and costing about the same, I assumed the Spark would provide choppy, nervous results with anything but the calmest, clearest conditions.

After a couple months with a test unit, my tune has changed. Blustery day trial flights at the coast didn't result in the Spark cartwheeling into the ocean, but instead produced sharp, locked-in video footage even as the controller screen blinked a red "high winds" warning. I've flown the little thing to roughly 1,500 feet away — the distance where I couldn't really spot it in the sky — but still maintained video and control transmissions. (It's advertised to go four times that distance, but I don't know that the battery life, a relatively short 10 minutes in real-world use, will take it that far, and definitely not there-and-back.) The forward-facing obstacle avoidance feature

works. And it is so crazy compact, it adds no discernible weight to my backpack even with its carrying case.

There are a few things I'd like to see improved or added, however. The follow-me modes feel like magic when they work, but when moving quickly on my skateboard at the park, they lose detection over half the time. There isn't a waypoint planning function like the bigger drones have, which could hinder mapping and automated search-and-rescue. The camera resolution maxes at 1080p. And that nice hard-foam case frustratingly leaves out a space for the separate controller (which you'll definitely want to get). I felt odd carrying it loose in my backpack.

Ultimately, I'm a hobbyist, not a professional drone flyer, which is something the Spark has made me realize and even embrace. It's now hard for me to look at my larger drones and feel good about their bulk. Anytime I need to get airborne, I just want to fly the Spark instead.

–*Mike Senese*

Hep Svadja

MU SPACEBOT

$60 morpx.com

The MU SpaceBot by Morpx is an interesting addition to the educational robot landscape. Unlike most of the robotic kits I see in this genre, it doesn't use locomotion as its main feature. Instead, the bot consists entirely of a 2-axis gimbal and a camera-equipped brain that allows it to see faces. Wrap that up in a simple wooden body and you have a robot that makes eye contact.

The ability to track faces really does make this robot stand out in a crowd. There's something very captivating about moving and having it follow you with its "eyes," quipping little pre-recorded messages.

After playing with the pre-set actions — face tracking and musical keyboard — you can jump into programming the SpaceBot yourself. This is done via an Android or iOS app that works with simple blocks for your actions, much like the Scratch programming environment. Telling it to look for a face and then say something, for example, took me about 30 seconds.

I really enjoyed not only the build, which was extremely easy, but also the customization of this robot. It came with blank adhesive panels for the body so you can make it look however you want. I had trouble connecting to mine over USB due to my phone's limitations, so I highly recommend getting the Bluetooth model ($69) for easier connectivity. This problem did allow me to test customer support though, which responded within a few hours. –*Caleb Kraft*

LJD61UP KEYBOARD KIT

Starting at $105 1upkeyboards.com

As the primary tool for most of us to access our computers, keyboards dominate a large portion of our days. While most are underwhelming, the kits from 1UP Keyboards aim to place a high quality device under your fingertips.

Made by keyboard enthusiasts for keyboard enthusiasts, these kits allow for an incredible amount of customization — some just for looks, some for the actual feel. Do you like your keys click-y or smooth? Should your keyboard be heavy-duty to hold up to your pounding, or light to travel with you? How about changing the colors of every part of your keyboard? This is all possible.

The keyboard I built was a stainless steel model with click-y mechanical keys. The GH60 PCB makes soldering easy, although it can be tricky lining up the keys to keep them all nice and straight. This kit takes a good bit of soldering, so be ready for it.

I love the feel of the keyboard — going back to my laptop's keys just felt flimsy after using the 1UP. If you spend your days typing away, give the 1UP a try. –*Matt Stultz*

AWK-105 ANALOG VOLTMETER CLOCK

$99 awkwardengineer.com

While walking through World Maker Faire this past fall, I spotted a booth with people lined up watching twitching analog voltmeters. As I stepped closer I realized they were clocks and I was instantly excited. I love clocks and find it fascinating all the ways we have created to represent time.

The Analog Voltmeter Clock from Awkward Engineer has two analog voltmeters on the front. These are the old school, needle-style meters that used to adorn almost every piece of electronics equipment before the advent of LEDs and digital electronics. One meter reads the hours, the other the minutes. The time is set and the dials are calibrated using two industrial-looking knobs on the top of the device. The aesthetics of this clock feel like it should be in a nuclear bunker during the Cold War. The warble mode adds some drama by making the dials hop periodically instead of being locked in on the time.

If you are looking for a timepiece for your shop, lab, office, or den, the fun of the Analog Voltmeter Clock will fit right in. –*Matt Stultz*

SUCKIT DUST BOOT **$80** suckitdustboot.com

Hobby CNC routers are wonderful machines to have in your life. They make beautiful, accurate carvings from wood, plastic, and many other materials. They also make a tremendous amount of noise and dust. The Suckit dust boot can help.

Available for both the X-Carve and the Shapeoko 3, the Suckit is a z-height independent dust collection system. This means you can set the height of the dust boot so that it stays level with the top surface of your material regardless of depth. The dust boot and support arms are machined acrylic, and the magnetically secured dust boot slides out easily to make bit changes and reinstallation a snap. The arms are positioned using thumbscrews to lock them in place, and the entire mechanism is attached to the z-gantry with just four Allen screws — the whole assembly can be removed in a few minutes if needed.

What's best is that the dust boot almost completely evacuates not only the fine particulate that's dangerous to breathe, but also the heavier matter that makes a mess of your workspace. The Suckit only accepts 2.5" dust collection hoses, so if you are using a smaller shop vac with a 1.25" hose you'll need to come up with an adapter to fit it. Nevertheless, the Suckit dust boot is a solid investment and a welcome addition to any CNC workshop. –*Tyler Winegarner*

BOOKS

DOT JOURNALING —
A PRACTICAL GUIDE
By Rachel Wilkerson Miller
$13 theexperimentpublishing.com

I had no clue dot journaling was even a thing — that's why I decided to read this book. Dot journaling takes its name from the grid-like series of dots that cover each page, like graph paper but simpler. With more versatility that regular ruled pages, Rachel Wilkerson Miller paints a clear picture of the many, many possibilities that all these dots can create. She gives beautiful examples in stunning detail, which will inspire you to practice them for yourself.

The best part is that she makes it a point to remind you that these are just suggestions and you are completely free to take what you will or leave it completely. I was so inspired that I decided to try this whole dot journaling thing out. I'm now five pages in and I love it. The flexibility is wonderful and I now have a place for my agenda items, planning thoughts, journal entries, and random things all neatly contained in one book. –*Jazmine Livingston*

PAINT THIS BOOK! WATERCOLOR FOR THE ARTISTICALLY UNDISCOVERED
By Thacher Hurd and John Cassidy
$25 theexperimentpublishing.com

I am *very* artistically undiscovered, and this book was absolutely meant for folks like me. It has just enough explanation in the beginning and very concise directions throughout. Plenty of tips and plenty of room for you to do your own thing. I especially like the "do whatever you want to get acquainted with art" in the very beginning before any technique is introduced. It definitely helps the artistically timid (like myself) feel better about art when we are explicitly told to do whatever and it's totally OK. –*Jazmine Livingston*

Make: *New & Notable*

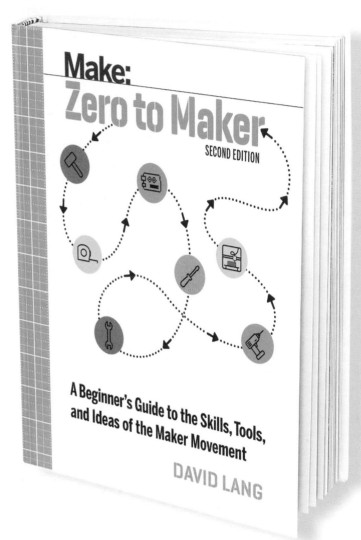

ZERO TO MAKER, SECOND EDITION
By David Lang $20

"David Lang ... teaches us that being a maker means far more than being able to pick up a soldering iron or run a 3D printer. It means opening a world of adventure, in which you aren't waiting for others to create experiences for you. Give yourself the power to explore your own dreams, just like he did." —Tim O'Reilly

MAKE: HIGH-POWER ROCKETS
By Mike Westerfield $35

Starting with an overview of mid- and high-power rocketry, readers will start out making rockets with F and G engines, and move on up to H engines.

EASY ELECTRONICS
By Charles Platt $10

This is the simplest, quickest, least technical, most affordable introduction to basic electronics, *ever*. No tools — not even a screwdriver — are necessary.

Recent Releases

Getting Started with Soldering
By Marc de Vinck **$20**

While new makers and experimenters will learn the core principles of soldering, more advanced makers will enjoy this as an excellent reference and resource book.

Getting Started with the BBC micro:bit
By Wolfram Donat **$20**

Learn all about this deceptively powerful pocket-sized computer.

mBot for Makers
By Andrew Carle and Rick Schertle **$20**

Perfect for non-technical parents, kids, and teachers who want to start with a robust robotics platform and then take it to the next level.

Fabric and Fiber Inventions
By Kathy Ceceri **$20**

Create and discover your own fabric inventions as you make wearables, playthings, and decorative items using the textile arts.

BROTHER SE625

Written by Matt Stultz

The addition of color touchscreen control makes a great new line of machines even better

THE SE625 MAKES PUTTING YOUR LOGOS AND DESIGNS ON FABRIC INCREDIBLY SIMPLE. Part of Brother's new line of combo machines, it makes some nice improvements on the popular SE4XX line. The 625 features a bright, relatively high-resolution, full-color touchscreen that allows you to see what your design will look like before producing. There are still a few important buttons on the front, but most controls are now handled with the touchscreen.

DESIGNS BUILT RIGHT IN

There are more than 80 built-in patterns and nine fonts. More patterns are available on an included CD, and you can create your own and copy them over using a USB stick. You will need special embroidery software to create your own designs that can be moved to the machine, but this unfortunately comes at a hefty extra cost. A free trial of Brother's PE Design is available though, so you can at least get started.

My first few attempts at getting the machine to embroider a pattern ended in failures, but I just adjusted the thread tension and it started humming along, creating beautiful patterns before my eyes. Multicolor jobs are not a problem. The machine will cut off the thread on its own and pause so you can swap out the thread for the next color.

THREADING THE NEEDLE

It may be silly, but one of my favorite features was the self-threading button. I've used plenty of sewing machines in the past, and we all know that getting the thread through the tiny hole in the needle can be tricky. The SE625 solves the problem with just the press of a lever, threading the needle perfectly each time.

EMBROIDER ALL THE THINGS

I run a hackerspace, and one thing we regularly do is put our logo on things to get ready for events. We make our own T-shirts, carve and laser cut signs, and stick vinyl logos to almost everything. With the help of an embroidery machine like the SE625, you can take the branding of your organization up a notch by creating more professional-looking shirts and fabric items. 🔊

- **WEBSITE** brother-usa.com
- **TYPE OF MACHINE** Embroidery and sewing
- **MANUFACTURER** Brother
- **BASE PRICE** $360
- **BUILD VOLUME** 102×102mm
- **ACCESSORIES INCLUDED AT BASE PRICE** Accessory feet: buttonhole foot, zipper foot, zigzag foot, button sewing foot, overcasting foot, blind stitch foot, monogramming foot, embroidery foot. CD of 200 embroidery designs, accessory pouch, bobbins (4), seam ripper, needle set, ball point needle, twin needle, cleaning brush, eyelet punch, screwdriver, spool caps (3), prewound bobbins with embroidery thread (3), extra spool pin, power cord, and operation manual.
- **ADDITIONAL ACCESSORIES PROVIDED FOR TESTING** None
- **WORK UNTETHERED?** Yes, internal memory, USB
- **ONBOARD CONTROLS?** Yes, full-color LCD touchscreen with buttons
- **OS** Windows for software. Designs can be transferred from a Mac as well.
- **OPEN SOFTWARE?** No
- **OPEN HARDWARE?** No

TAKE THE BRANDING OF YOUR ORGANIZATION UP A NOTCH WITH THE HELP OF AN EMBROIDERY MACHINE

PRO TIPS

Embroidery software can be more expensive than the machines themselves. Don't worry, there are some lower cost choices available. Read our full guide to software options at makezine.com/go/embroidery-software

WHY TO BUY

The bright, full-color touchscreen is a great upgrade to the popular SE4XX family of embroidery machines, making the SE625 easier to use and easier to preview what you are going to get in your final product.

Hep Svadja

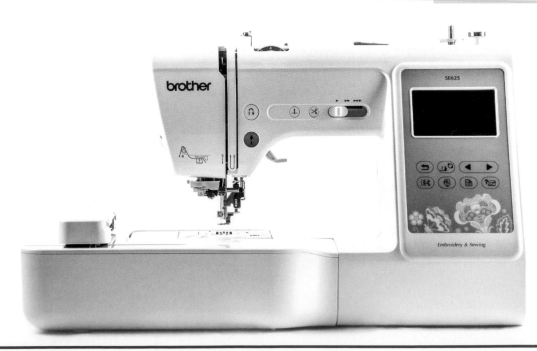

SHOW & TELL

Dazzling projects from inventive makers like you

Sharing what you've made is half the joy of making! To be featured here, share your projects with us on Maker Share!

1. **Wes Swain** immortalized his love for retro games within this one-of-a-kind cement Thwomp.

4. **Esya Rachma** came up with this ingenious solution to hang-dry clothing during sporadic weather conditions.

7. This amazing player was created by **Alain Mauer** so his son who has autism could watch media on his own.

2. This simple RFID-controlled MP3 player was created so **Martin Jacob** could give his 4-year-old musical agency.

5. Not only does this clock tower look super cool, but **5 Volts** designed it so that he's the only one who can read the time.

8. Tired of hauling around backpacks, **Tanner Packham** decided to give his new rolling bag a motorized upgrade.

3. Looking for a way to engage passers-by, **Matthew Dalton** and **Markus Schilling** designed these data pillars to track movement, wind, and sunlight.

6. This weather display can be embedded in just about any project, or simly placed on a desk, which is **Alex Wulff's** current setup.

9. A collaborative project for People in Need Cambodia, this flood warning system by **Rob Ryan-Silva** is saving lives.